D1736333

1/21/74

# Electronics in the Life Sciences

STEPHEN YOUNG

*Lecturer in Zoology*
*Imperial College, London*

A HALSTED PRESS BOOK

JOHN WILEY & SONS
New York

First published in the United Kingdom 1973 by
The Macmillan Press

Published in the U.S.A. by
Halsted Press, a Division of
John Wiley & Sons, Inc. New York

Printed by photolithography and bound in Great Britain

**Library of Congress Cataloging in Publication Data**

Young, Stephen.
Electronics in the life sciences.

"A Halsted Press book."
1. Electronics in biology. I. Title.
QH324.Y68 1973 574'.072 73–8083
ISBN 0–470–97943–7

# Preface

Complex electronic instrumentation is rapidly becoming ubiquitous in biological laboratories. Animal nervous systems use small electrical signals to transfer information and to compute strategies, and those studying them need electronic amplifiers and dataloggers. Many biochemical and physiological data are most readily measured and recorded if they can first be reliably transformed into electrical signals by transducers. Temperature changes, illumination changes, and movements of an animal are all typical examples. Moreover it is often necessary to maintain the environment of an experimental animal, plant, or biochemical system in conditions of constant and measured temperature, humidity and illumination, or to provide it with stimuli changing in a stereotyped fashion. All these require electronic servomechanism control systems.

Electronic event recognisers coupled to datalogging systems have extended the range of experimentation to situations in which the rates at which events occur are too fast or too slow for the unaided human observer: if signals arrive at more than two per second human observers soon begin to make errors, and similarly if the gaps between signals are of more than two or three minutes the observer simply drops off to sleep.

I have the strong feeling that these complex electronic systems, however indispensable, are disliked and distrusted by many biologists. There is a considerable nostalgia for the paper, pencil, string and sealing wax phase even among those too young to have any personal experience of it, and a tendency to treat the electronics with alternating contempt (a good thump when it doesn't work) and exaggerated respect (summoning the firm's representative every time a minor adjustment to a knob on the front panel is needed). I think it's probably worth making clear at this point that no well-designed piece of electronics can ever be damaged by any combination of settings of the knobs on its front panel, and that the best way of finding out the functions of the control knobs in behavioural rather than electronic terms (knobs tend to be labelled in terms which tell you what they do to circuits rather than how this affects the final output of the machine) is to set the thing going and to make small adjustments to each control in turn and to watch what happens.

But even those who have satisfactorily come to terms with their instruments, and who have opted for the strategy of buying as much as possible of their equipment readymade still tend to come up against difficult electronic problems when it comes to connecting the various boxes together ('interfacing'). It is frequently necessary to attenuate or amplify voltages, to change the length of pulses, and so forth.

Hence many students and research workers in biology and medicine now require a working knowledge of electronics, and I have tried to arrange this book to cater for a wide variety of their needs. The student who requires a textbook to supplement a course, or the research worker with little or no previous experience in electronics, can simply read the book straight through. The first four chapters supply a minimum of electronic theory, and the rest of the book deals with commonly used circuits, arranged in order of increasing conceptual difficulty. For the more experienced user, the electronic ideas are always introduced in the context of the practical applications listed as chapter headings, and so he should be able to find both the theoretical and practical information needed to build any given circuit within a few pages of text. However, neither of the superimposed arrangements corresponds to a rigorous ordering of circuits on an electronic basis, and so a schematic table of contents is provided which classifies every type of circuit in the book in a rational electronic scheme.

In addition to providing this basic information, I hope that I have also managed to demystify much of the electronic terminology which has crept insidiously into so much biological model building and theory forming. Everybody seems to be talking about positive and negative feedback, AC coupling, relaxation oscillators and the like, often without very much idea about the very precise engineering definitions of these terms, and certainly without the wealth of technological examples which have made engineering examples so fruitful in the past.

I have actually built all the circuits in this book, and they do all work. However it is difficult to offer an unconditional guarantee, and I should be very interested to hear of any problems. It is very hard to remember the exact provenance of all the circuits, and I must apologise to anyone whose 'original' circuit I have unwittingly borrowed. The only two electronics books I use frequently are:

T. D. Towers *Elements of Transistor Pulse Circuits*
P. E. K. Donaldson *Electronic Apparatus for Biological Research*

All my electronic thinking has been influenced by Steve Salter's teaching, and my approaches to instrumentation problems stem largely from discussions and co-operation with Anthony Downing, Ian Fosbrooke, Richard Gregory, Jim Howe, and John Moorhouse.

Stephen Young

# Contents

# Circuit List

## PASSIVE (NO POWER SUPPLY)

## ACTIVE

### Amplifiers and Integrators

## ACTIVE

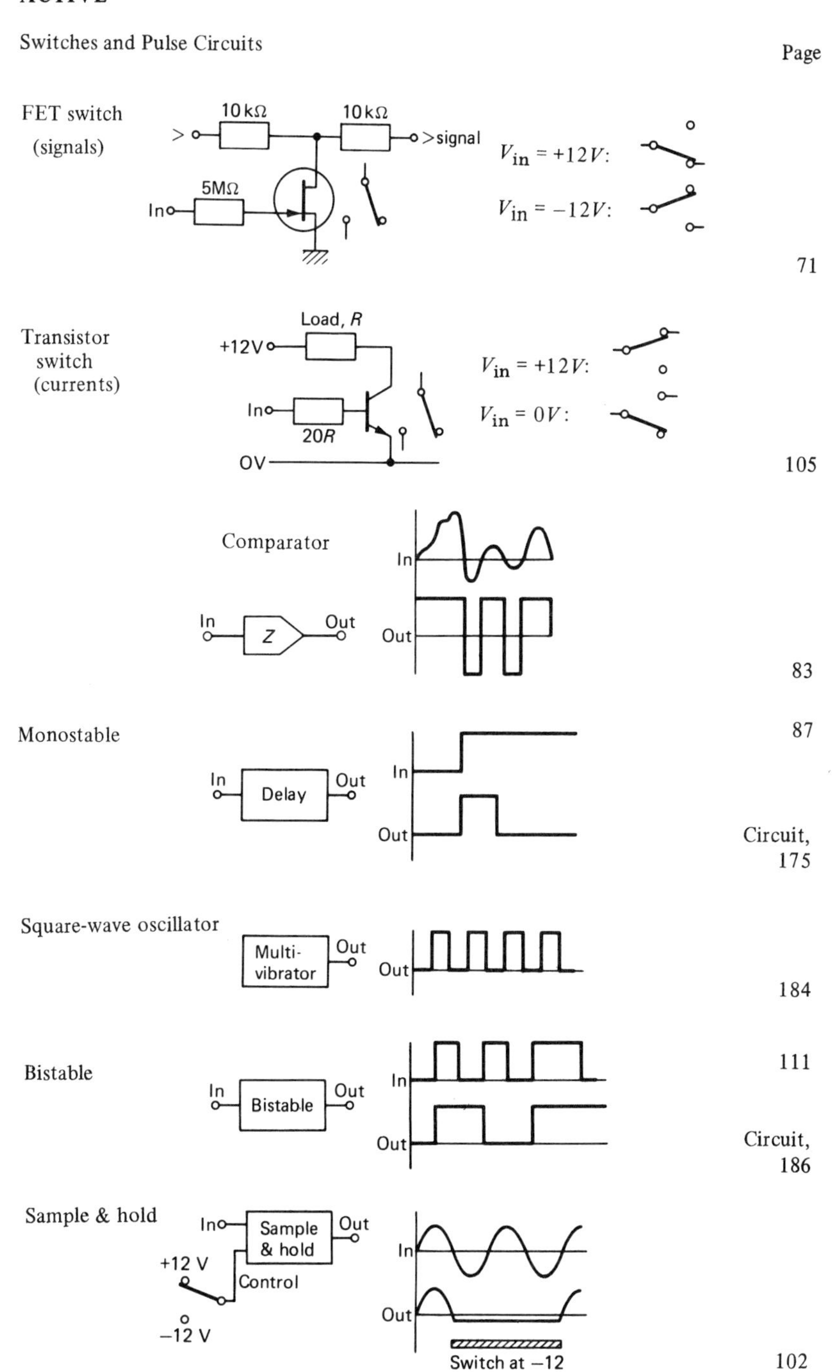

Switches and Pulse Circuits
Page
FET switch
(signals)
10 kΩ
10 kΩ
>signal
5MΩ
In
Vin = +12V:
Vin = −12V:
71
Transistor
switch
(currents)
Load, R
+12V
In
20R
0V
Vin = +12V:
Vin = 0V:
105
Comparator
In
Out
Z
83
Monostable
In
Delay
Out
87
Circuit,
175
Square-wave oscillator
Multi-
vibrator
Out
184
Bistable
In
Bistable
Out
111
Circuit,
186
Sample & hold
In
Sample
& hold
Out
+12 V
Control
−12 V
Switch at −12
102

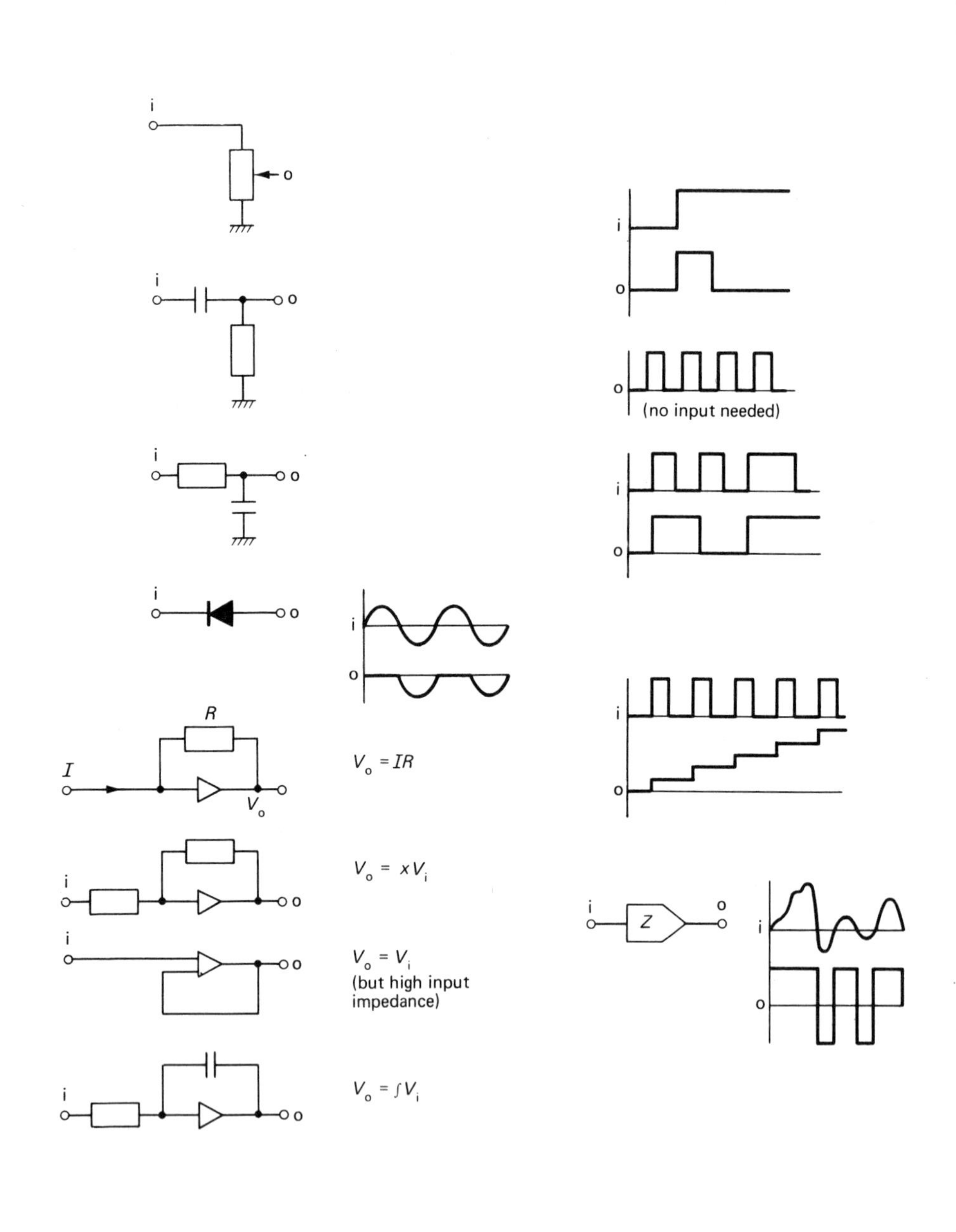

i
o
i
o
i
o
i
o
o
(no input needed)
i
o
i
o
R
I
$V_o$
$V_o = IR$
i
o
i
o
$V_o = xV_i$
i
o
$V_o = V_i$
(but high input impedance)
i
o
$V_o = \int V_i$
i
o
i
o
Z
i
o

# 1 The Multimeter

The multimeter is the tool for measuring steady electrical quantities. The most common sort is the avometer which is capable of measuring amps, volts or ohms, depending on the position of the rotary switches on the front. It is necessary to decide which is the sensible scale to read under the meter needle, and to place the instrument appropriately in the circuit.

## 1.1 AMPS

The most basic electrical quantity is the amp, which measures electric current, or the rate of flow of electrons through the various wires constituting a circuit. A current of one amp means that $6.038 \times 10^{18}$ electrons per second are flowing along the wire in question. To measure the current in a circuit, it is necessary to break the circuit and insert the meter (figure 1.1).

The meter uses the fact that an electric current passing through a coil of wire generates a magnetic field whose strength depends on the strength of the current.

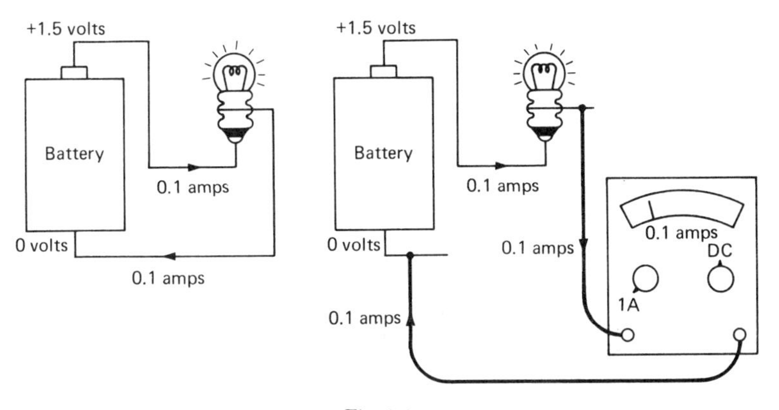


Fig. 1.1

The coil is attached to the meter needle, and the stronger the current the further the needle moves against the spring pulling it back.

Unfortunately, currents flow round the circuit the opposite way from electrons. This has arisen because the decision as to what should be positive and what should be negative (currents always flow from positive to negative) was made before electrons were discovered, so which pole of the battery should be called positive and which negative was purely a matter of convention. It is just unlucky that when an actual particle to carry the electric charge round a circuit finally came to light it should prove, in terms of the established convention, to be 'negatively charged' and hence to move always in the opposite direction to the conventional currents.

## 1.2 VOLTS

It is clear that the current flow in the circuit shown is accompanied by work being done, the small bulb is giving out light and heat, and the chemicals in the battery are changing from high energy to low energy states; in other words the battery is running down. This process does not use up any electrons – the current flowing into the battery is just the same as the current flowing out of it; indeed the current at any point in a closed circuit without any branches always has the same value. But the potentialities of the electrons to do work do change as they pass round the circuit, and these changes are represented by voltage differences between different points of the circuit.

The torch bulb in figure 1.2 has a voltage difference of 1.5 volts maintained across it by the battery. To measure a voltage, it is not necessary to break into the

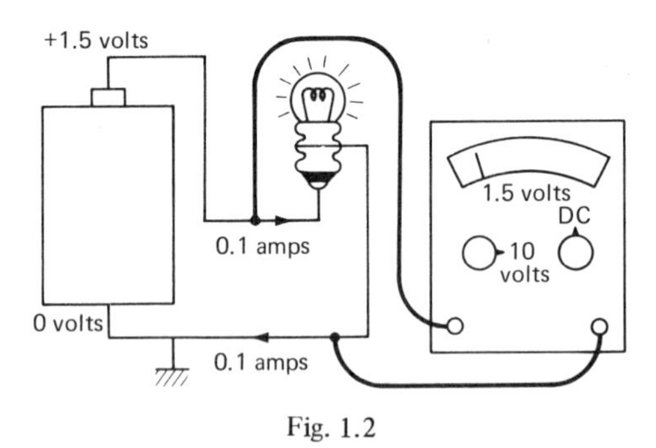


Fig. 1.2

circuit. The voltage differences between various parts of the circuit are the important quantities and it does not matter which part of a circuit we decide to label '0 volts' provided the direction and size of the voltage differences are correctly labelled. Figure 1.3 is an equally valid representation of this same circuit. It is sometimes useful to connect one point of a circuit to 'ground' or 'earth' (the third pin of the British mains supply, which leads to a water pipe buried in the earth.) This grounded point is always labelled '0 volts', and other voltages in the circuit are made positive or negative accordingly. The symbol ⏚, or sometimes ⏚ in older books, is used to indicate the point at which the circuit is earthed or grounded.

The useful property of batteries and the electronic power supplies driven from the mains which frequently replace them is that they maintain a constant voltage across external circuits which consume differing amounts of power. For instance if a mains voltage bulb is connected across the 1.5 volt battery used above, the voltage across it is still 1.5 volts, though the bulb does not light up and consumes scarcely any current (figure 1.4).

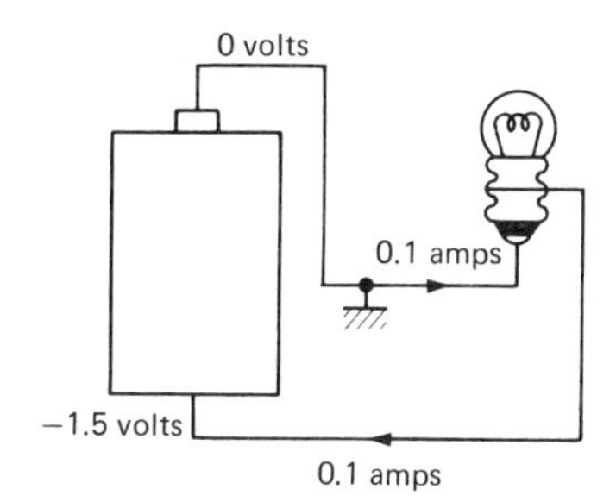


Fig. 1.3

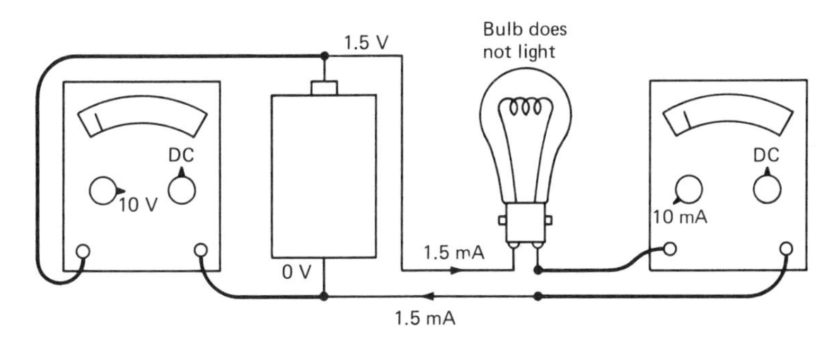


Fig. 1.4

Milliamps (mA), thousandths of an amp, are used as a measure of small currents. The power consumed in a circuit is measured in watts

$$(\text{Volts across circuit}) \times (\text{Current flowing through it}) = \text{Power}$$

$$\text{Volts} \times \text{Amps} = \text{Watts}$$

So the torch bulb consumes

$$1.5 \times 0.1 = 0.15 \text{ watts}$$

Watts are directly convertible into the mechanical units of power – there are 746 watts in one horsepower.

## 1.3 OHMS

The mains bulb in the above example passes less current and consumes less power than the torch bulb when both are connected to 1.5 volt batteries because its *resistance* to the passage of electric current is greater.

This electrical resistance is the third of the important electric quantities and is measured in ohms (Ω).

$$\text{Ohms} = \frac{\text{(Volts across circuit)}}{\text{(Amps flowing through it)}}$$

This relationship is always called Ohm's Law, although it is really a definition rather than a law. The torch bulb hence has a resistance of 1.5/0.1 = 15 ohms and the main bulbs $1.5/(1.5 \times 10^{-3}) = 10^3 = 1000$ ohms. 1000 ohms is called a kilohm (kΩ).

In the case of the light bulbs, the electrical resistance depends on the voltage applied to them – when they are lit up they have a much higher resistance than when the wire filament is cold. But it is possible to manufacture components which have a constant resistance whatever the voltage applied to them, that is if the voltage across them is doubled, the current flowing through will exactly double too. These components are called resistors and they have a striped appearance (figure 1.5). These stripes indicate their value and how accurately they have been constructed. This practice of coding seems irrational at first since it would seem simpler if the

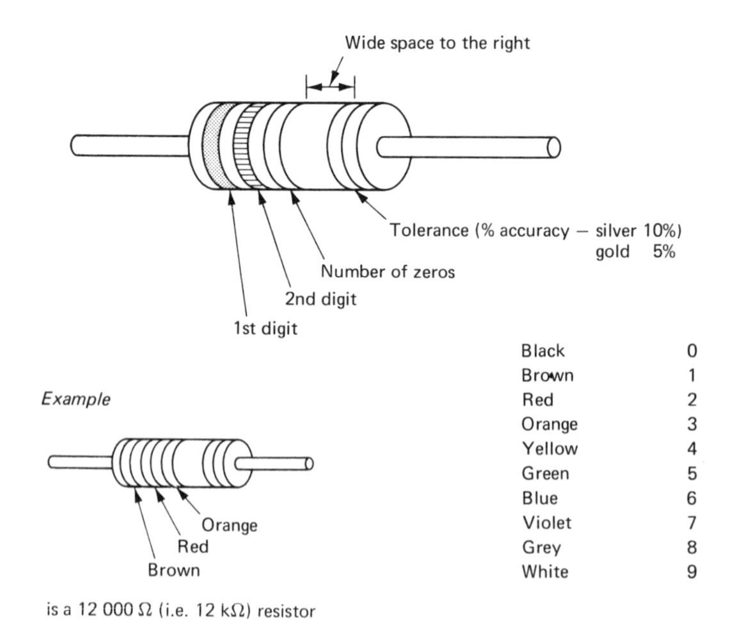


Fig. 1.5

values were just written on in numbers, but it does have the advantage that you can always read the value of a resistor when it has been soldered onto a circuit board crammed against other components. The symbol for a resistor used in circuit diagrams is –▭– although –/\/\/\/\–is sometimes used.

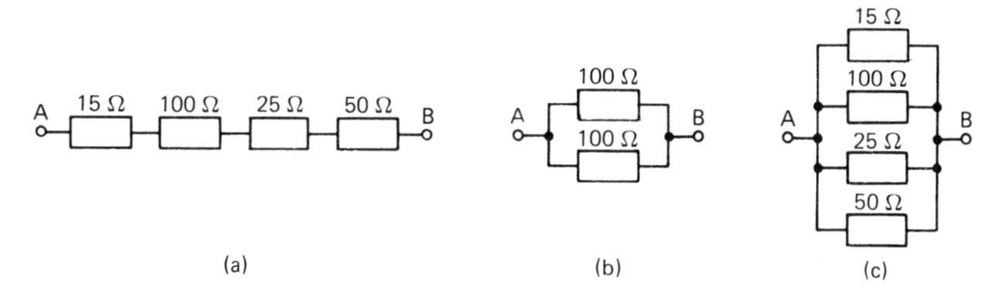


Fig. 1.6

If resistors are connected in series, their values add (figure 1.6*a*)

$$\text{Total resistance between A and B} = 15 + 100 + 25 + 50 = 190\ \Omega$$

For parallel connection however (figure 1.6(b)) total resistance between A and B = 50 Ω because

$$\frac{1}{R_{\text{total}}} = \frac{1}{100} + \frac{1}{100}$$

and so (figure 1.6(c))

$$\frac{1}{R_{\text{total}}} = \frac{1}{15} + \frac{1}{100} + \frac{1}{25} + \frac{1}{50}$$

$$= \frac{20 + 3 + 12 + 6}{300} = \frac{41}{300}$$

$$R_{\text{total}} = \frac{300}{41} = 7.5\ \Omega$$

Voltages down a chain of resistors are found by simple proportion (figure 1.7). An avometer can be used to measure the resistance of a component by setting the controls to an ohms range and clipping the meter leads onto each end of the component. A little care is needed because the ohms scales are backwards, with the zero at the right-hand end, and high values to the left. The meter in fact measures how much current from an internal battery can get through the component being measured, and the higher its resistance the lower the current. A knob labelled 'zero ohms' is used to correct for the resistance of the meter leads and any falling off in the battery. This is done by clipping the leads directly together, and adjusting the knob till the pointer is over the zero calibration.

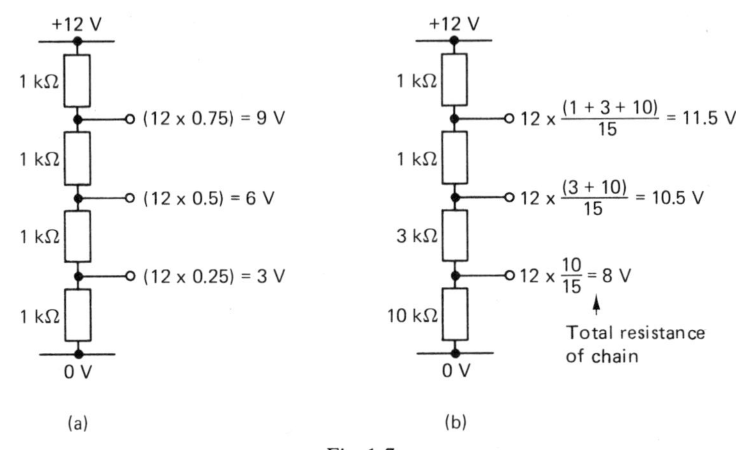


Fig. 1.7

The traditional way to make a more accurate resistance measurement is to match the component being measured with a resistance box, which typically contains nine each of accurately made 1, 10, 100 and 1000 ohm resistors, connected so that any resistance between 1 and 9999 ohms can be selected according to the position of four rotary switches. The matching involves making a voltage divider from the component being measured and from a known resistor ($R_1$) (figure 1.8).

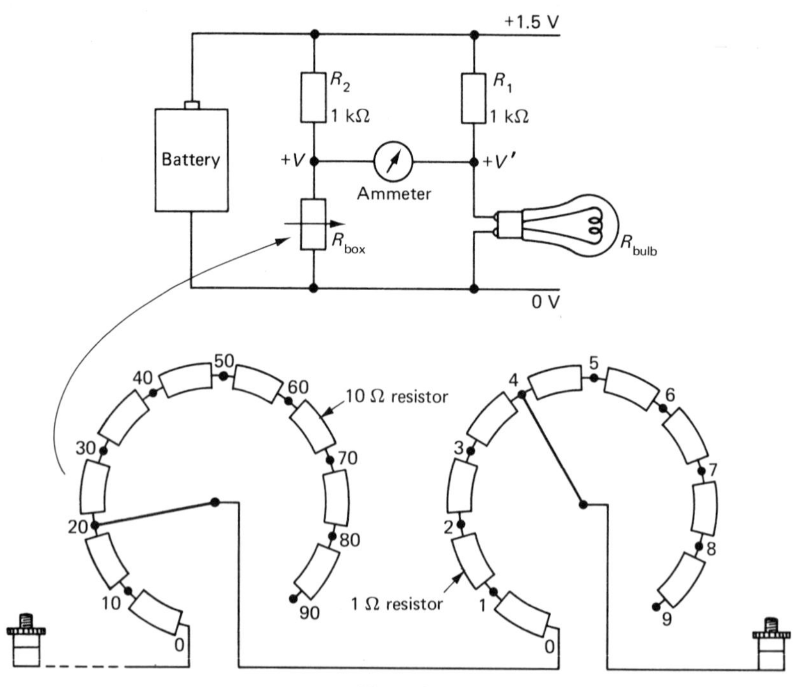


Fig. 1.8

A second known resistor ($R_2$) and the box make up a second voltage divider. A sensitive ammeter connects the centres of the two dividers, and the box's resistance is adjusted till the ammeter reads zero amps. No current flows only between points at the same voltage, so $V = V'$. Hence the ratios of the two voltage dividers are the same:

$$\frac{R_2}{R_{box}} = \frac{R_1}{R_{bulb}} \text{ and, since } R_2 = R_1 = 1 \text{ k}\Omega$$

$$R_{box} = R_{bulb}$$

This circuit is called the Wheatstone bridge, and the process of adjusting the box resistance till the ammeter reading is zero is called balancing the bridge. With the values shown it is only possible to balance the bridge if the resistor being measured is less than the largest box resistance of 9999 ohms. If it is bigger than this it is necessary to change $R_1$, say, to 10 k$\Omega$.

Then

$$\frac{1000}{R_{box}} = \frac{10\,000}{R_{unknown}}$$

and hence

$$R_{unknown} = 10\,R_{box}$$

When it is essential to pass only a very small current through the resistor being measured, use the method described on page 140.

## 1.4 AMMETERS AND VOLTMETERS

An ammeter should have a very low resistance compared with anything else in the circuit so that the voltage dropped across it is slight, and hence the interference with the behaviour of the circuit into which it is inserted is minimal.

The avometer consists of a very sensitive ammeter with 50 microamps ($\mu$A) – millionths of an amp – as full-scale deflection. If necessary, it can be made less sensitive by connecting small resistors across it (figure 1.9). If the resistance of the ammeter coil is 50 ohms, and that of the shunt resistor 12.5 ohms, then the current $I$ will be shared between the two components so that four times as much passes through the resistor as through the meter. The meter will now need a current of 250 microamps to give a full-scale deflection; in other words it has been made five times less sensitive. The switch on the front of an avometer changes the value of the shunt resistor as the different amps ranges are selected.

A voltmeter should have a very high resistance so that from the point of view of the rest of the circuit it does not appreciably alter the resistance of the component with which it is connected in parallel.

Voltmeters basically consist of an ammeter plus a series resistor (figure 1.10). The ammeter records the value of the current, $I$ amps, flowing from the bottom of the series resistor to earth. Since the series resistor has a value 1000 times greater than that of the meter coil, 99.9 per cent of the voltage $V$ which is being measured is applied to the resistor. So the current $I = V/50\,000$ amps the full-scale deflection of the meter, $I = 50$ microamps, will thus be produced by $50\,000 \times 50 \times 10^{-6} =$ 2.5 volts applied across the meter and its series resistor.

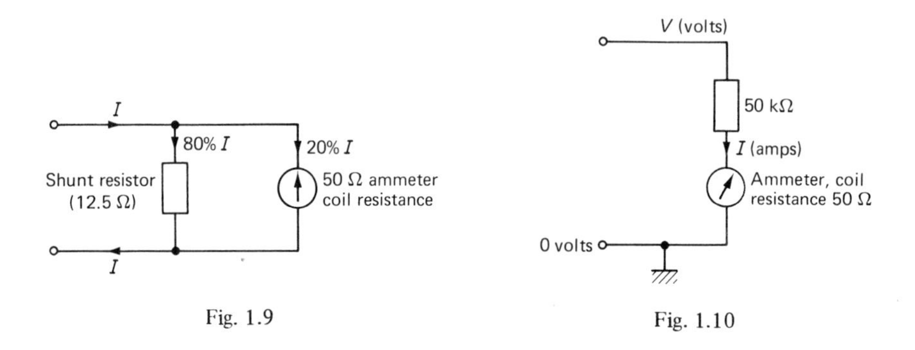


Fig. 1.9

Fig. 1.10

The voltage ranging switches on an avometer alter the value of this series resistor.

Many electronic circuits involve components of several million ohms (megohms – MΩ) resistance. In such circumstances any moving coil meter is going to give misleading readings since its resistance is so low that it will draw more current than any of the other components in the circuit. A safe rule is to use the avometer to measure voltages only across resistors of less than 5 kΩ; for higher resistance circuits an oscilloscope or digital voltmeter must be used.

## 1.5 AC RANGES

All the voltages and currents discussed so far have been DC (direct current) – they have been produced by batteries or special power supplies. The electricity mains are AC (alternating current) (figure 1.11). Clearly an AC supply is of constantly changing voltage; the voltage recorded on the avometer AC range is an average arranged so that an electric fire will give out the same amount of heat on an AC or DC 240 volt supply.

When the AC ranges of an avometer are selected, a component called a diode is connected internally in series with the meter (figure 1.12). This is a silicon or germanium device with the property of conducting electricity only in one direction, that shown by the arrow on the symbol (figure 1.13). The effect of this component on the AC input to the meter is to remove all the negative-going portions (figure 1.14).

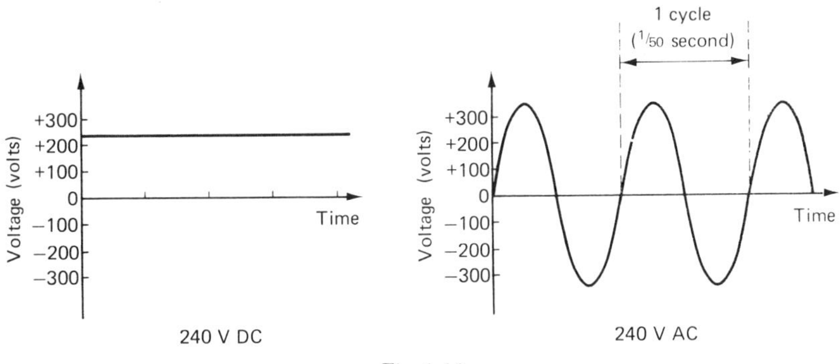


Fig. 1.11

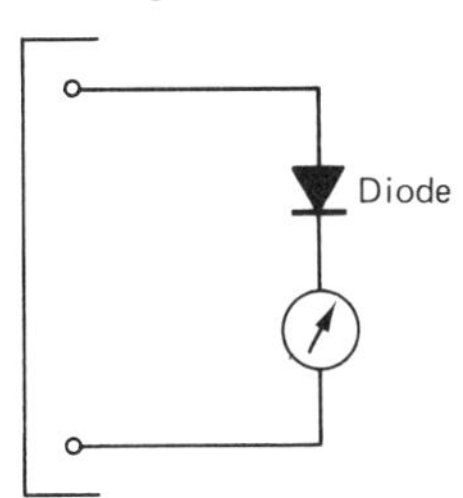


Fig. 1.12

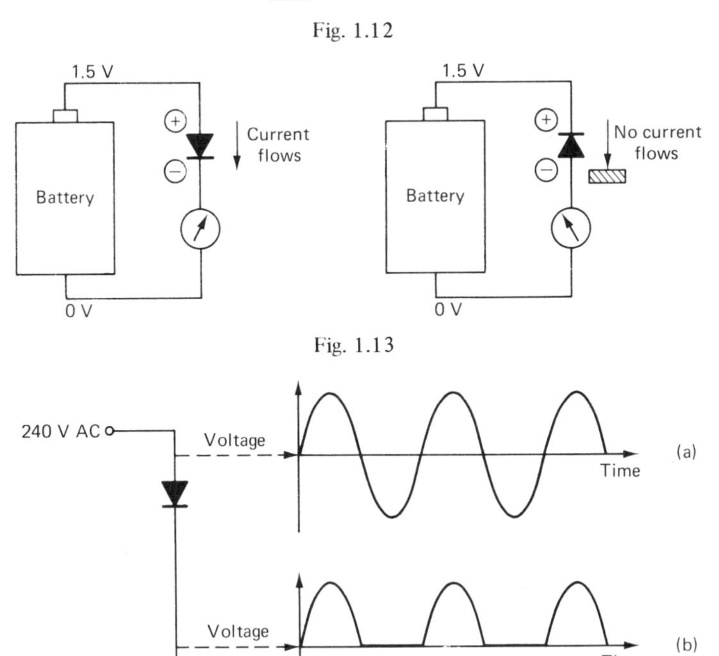


Fig. 1.13

Fig. 1.14

When a moving coil meter is subjected to a current alternating too fast for the coil following it to move, the reading obtained is an average value. In the case of the AC in (a), the average voltage is zero, since each positive-going peak is exactly cancelled by the following negative-going trough. After passing the diode ('rectification'), the removal of the troughs leads to a situation in which there is a nett positive average voltage, and hence the meter can respond with a reading proportional to the amplitude of the original AC voltage.

# 2 The Oscilloscope

The oscilloscope on the right in figure 2.1 is plotting a graph of voltage ($Y$-axis) against time ($X$-axis) for the repetitively changing electrical voltage (the 'signal') being output by the oscillator on the left. The use of the word signal to mean any train of electrical events produced by one device and which affect a second shows the way so much modern electronics has its roots in telegraphy.

## 2.1 SIGNAL SOURCES

### Oscillators

Oscillators are often used to supply inputs to circuits which are being developed or tested. They can be adjusted to mimic the output from the biological preparation or from any other electronic circuit which the device normally receives, and serve to localise any faults to the one circuit being examined. The oscillator in the photograph is called an audio oscillator because it operates in the range of frequencies used for sound recording. The oscillator controls alter the *frequency* and *amplitude* of its output. Frequency is the number of cycles per second, a cycle being the largest portion of signal you can get without it starting to repeat itself. The amplitude of a signal is the difference between the lowest (or most negative) and the highest (or most positive) voltage it contains (figure 2.2).

In this case the frequency control enables the output signal to be set to any frequency between 1.5 and 150 000 cycles per second. The unit of frequency is the Kilohertz (Hz): 1 hertz is equal to 1 cycle per second, and 1000 cycles per second is 1 kHz.

The amplitude control adjusts the amplitude from zero to 2.5 volts.

The shape of the signal can be made sinusoidal as shown (this follows the graph produced by plotting a table of sines), or square. (See figure 2.1, control labelled waveform.)

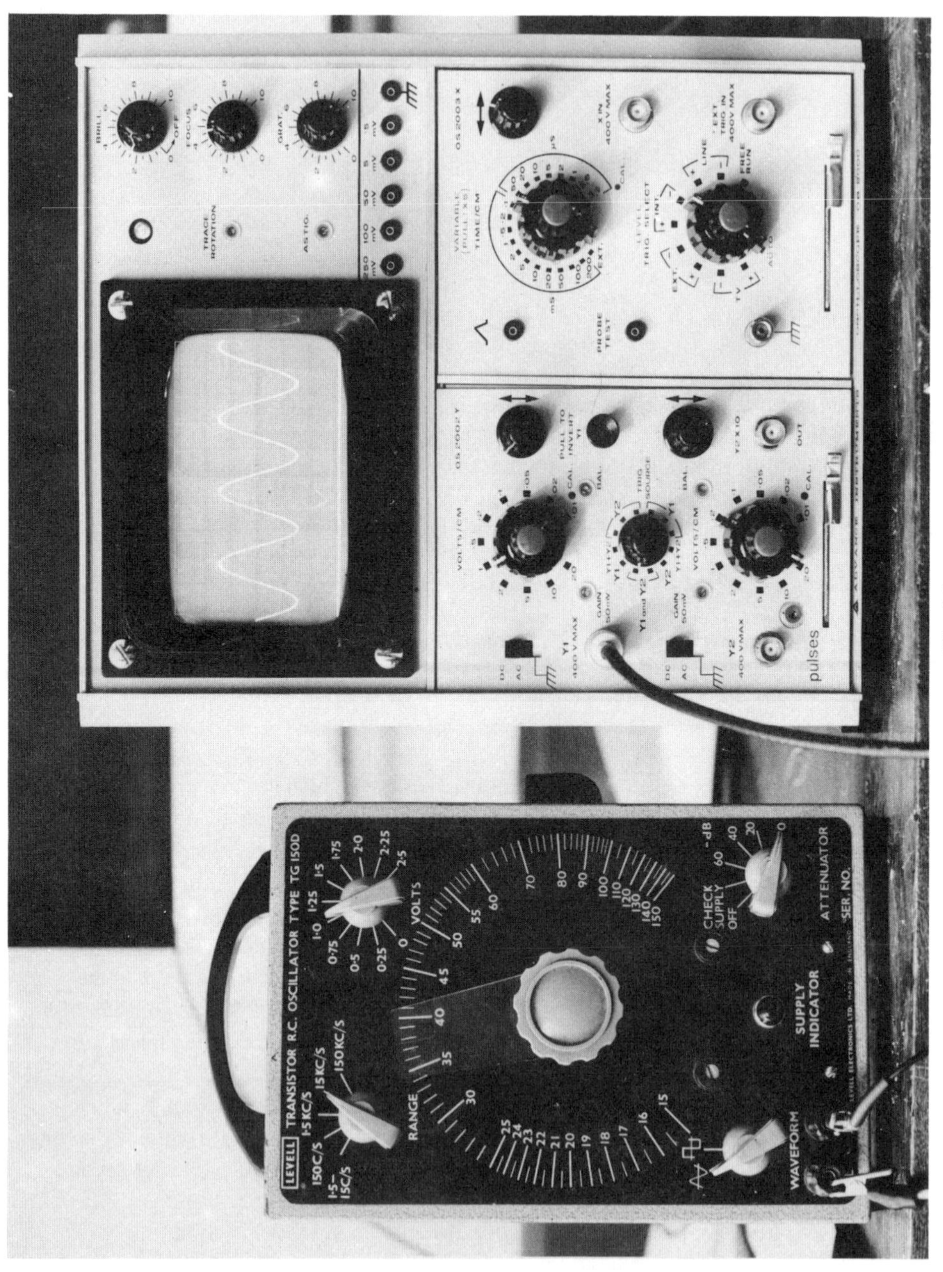
LEVELL TRANSISTOR R.C. OSCILLATOR TYPE TG 150D
RANGE
VOLTS
WAVEFORM
SUPPLY INDICATOR
CHECK SUPPLY OFF
ATTENUATOR
SER. NO.
TRACE ROTATION
ASTIG.
TIME/CM
VOLTS/CM
PROBE TEST
TRIG. SELECT
pulses

Fig. 2.1

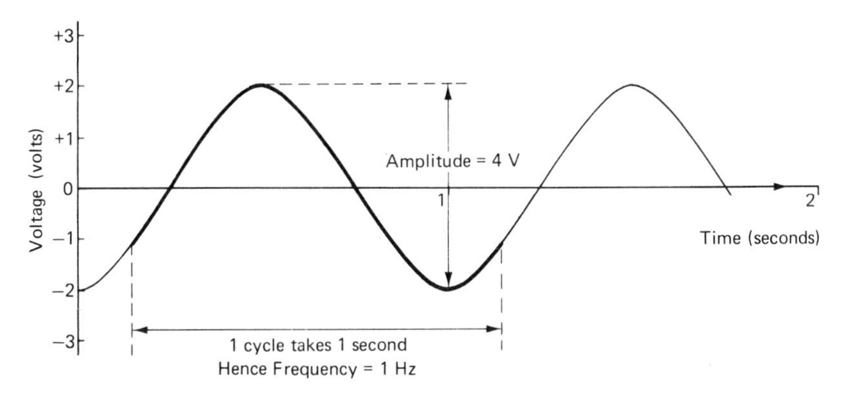


Fig. 2.2

**Decibels**

Oscillator controls are clear and self-explanatory, except for the ATTENUATOR (bottom right), used to reduce the output voltage in big steps. This is calibrated in decibels (dB), a confusing unit frequently used to describe the ratio between the input and output of an attenuator (a device whose output voltage is a constant fraction of its input voltage) or an amplifier (whose output is a constant multiple of its input).

$$\text{Decibels} = 10 \log_{10} \frac{(\text{Input power})}{(\text{Output power})}$$

The subscript 10 is the 'base' of the logarithms – logs to the base ten are those in ordinary tables.

Now power (watts) = $I$(amps) x $V$(volts) (see Chapter 1). For a resistor $I = V/R$, so

$$\text{Power} = \frac{V}{R} \times V = \frac{V^2}{R} \text{watts}$$

Thus

$$\text{Decibels} = 10 \log \left[ \frac{\dfrac{(\text{Input volts})^2}{R_{\text{input}}}}{\dfrac{(\text{Output volts})^2}{R_{\text{output}}}} \right]$$

If $R_{\text{in}} = R_{\text{out}}$, then

$$\text{Decibels} = 10 \log \left( \frac{V_{\text{in}}^2}{V_{\text{out}}^2} \right)$$

Now log (anything)$^2$ = 2 log (anything). Hence

$$\text{Decibels} = 20 \log \left(\frac{V_{\text{in}}}{V_{\text{out}}}\right)$$

Table 2.1 gives some examples of the application of this formula.

Table 2.1

| Dimunition of signal | | Attenuator switch setting (20 log ratio) |
|---|---|---|
| $\frac{1}{10}$ | (log 1/10 = –1) | –20 dB |
| $\frac{1}{100}$ | (log 1/100 = –2) | –40 dB |
| $\frac{1}{1000}$ | (log 1/1000 = –3) | –60 dB |

## 2.2 OSCILLOSCOPE CONTROLS

### *Y*-axis of display

It is very important to be able to adjust an oscilloscope quickly to produce a clear relevant display. It is worth checking that you can obtain the results illustrated below with an actual oscilloscope.

The *Y* controls (figure 2.3), on the left-hand panel, affect only the vertical axis of the graph being plotted on the screen. There are two sets of these controls because this oscilloscope has two beams, enabling two independent input voltages to be graphed simultaneously and thus compared. In the series of illustrations no input is connected to the second beam, which remains as a horizontal straight line at the bottom of the screen.

The *Y*-shift knob moves its trace bodily up and down the screen, that is it alters the vertical position of the origin of the graph. Its main use it to place the lower extreme edge of the trace against a graticule line to make it easier to read off voltages from the graph. It can also be used in cases where the input voltage has a steady component which would otherwise push the graph off the top or bottom of the screen.

The latter is a common cause of total absence of picture; in such a case, after making sure that the oscilloscope is in fact connected to the mains and switched on (pilot light? Illuminated graticule working?), first check that the *brightness* control is turned up far enough to produce a picture. Choose a brightness level high enough to see the picture comfortably, for modern phosphors (the inner coating on display

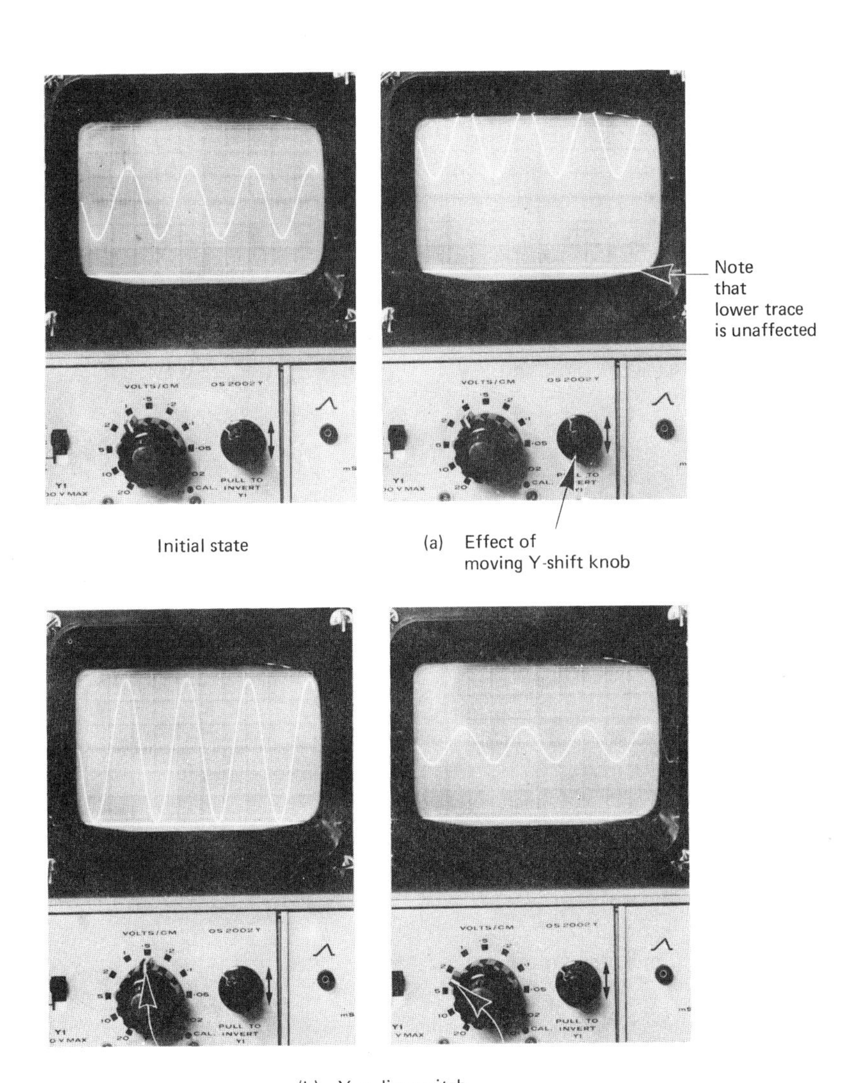


Initial state

(a) Effect of moving Y-shift knob

(b) Y-scaling switch

Fig. 2.3

tubes) are tough and there is no need to peer at dim traces for fear of burning the phosphor. Then twist the shift control back and forth, which should bring a trace into view.

Next adjust the *Y*-scaling switch (b) which controls the gain of the amplifier connecting the input socket to the display tube. This switch is calibrated with the number of volts each vertical centimetre of the graph represents on that particular setting, and turning the knob clockwise increases the gain of the amplifier, diminishes the number of volts each centimetre represents, and makes the height of the trace larger for an unchanging input. It is best to choose a setting which gives a trace occupying about half the height of the screen, to leave room for something else to be plotted with the second trace.

If the *Y*-scaling control is put at very much too sensitive a setting the 'no picture' syndrome can result, the tiny fraction of the middle of the picture appearing on the screen being too faint to show. It is hence always worth starting out with this control on an insensitive setting.

### *X*-axis

Next come the *X* controls on the right panel (figure 2.4). The *X*-shift (a) moves the display bodily sideways. It sets the position of the origin of the graph in the horizontal direction. This control is adjusted rather rarely, for we generally want to keep the display central on the screen. However it is worth checking in recalcitrant cases of 'no picture'.

The timebase repetition-rate control (b) calibrates the *X*-axis of the graph, the setting being the number of seconds each horizontal centimetre represents. (ms = milliseconds = thousandths of a second, μs = microseconds = millionths of a second.) The 'timebase' is the name given to the circuit which controls the horizontal position of the spot. Its operation can be seen most clearly if the knob is turned to one of the counterclockwise positions, with a large time interval per centimetre. The steady trace typical of fast timebase speeds can now be resolved by the eye into a spot which moves steadily from the left of the tube to the right, then instantaneously reappears at the left-hand end and starts again.

These slow settings are useful in watching slow moving irregular phenomena; for instance the voltage changes produced in a strain gauge attached to a locust's leg when the animal walks. The tube phosphor continues to glow for a little after the spot has passed over it, so a fleetingly visible record is maintained on the screen. Phosphors differ in their length of afterglow or 'persistence'. For biological work long persistence phosphors (e.g. type P7) are best.

As the timebase control is turned clockwise the speed of movement increases until the moving spot is seen as a line. Very fast speeds can cause the picture to become very faint if no high frequency signal is present, so it is best to start with a moderate speed.

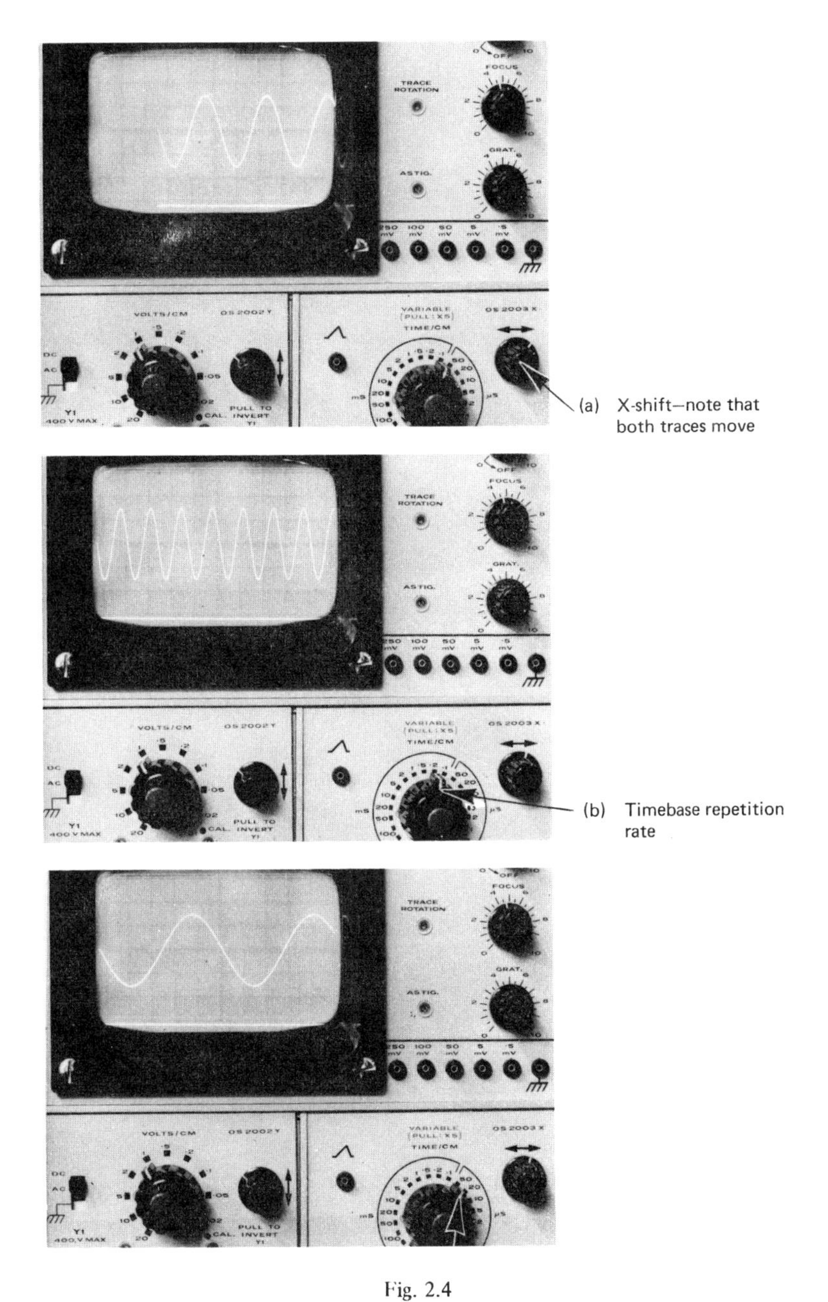
(a) X-shift—note that both traces move
(b) Timebase repetition rate

Fig. 2.4

**Timebase Triggering**

In order to produce a legible display at the faster timebase speeds the oscilloscope must 'trigger'. This means that each brief traverse of the spot across the screen must lie exactly on top of its predecessor, so that they add together to produce the steady picture of a repetitive input signal shown in the illustrations.

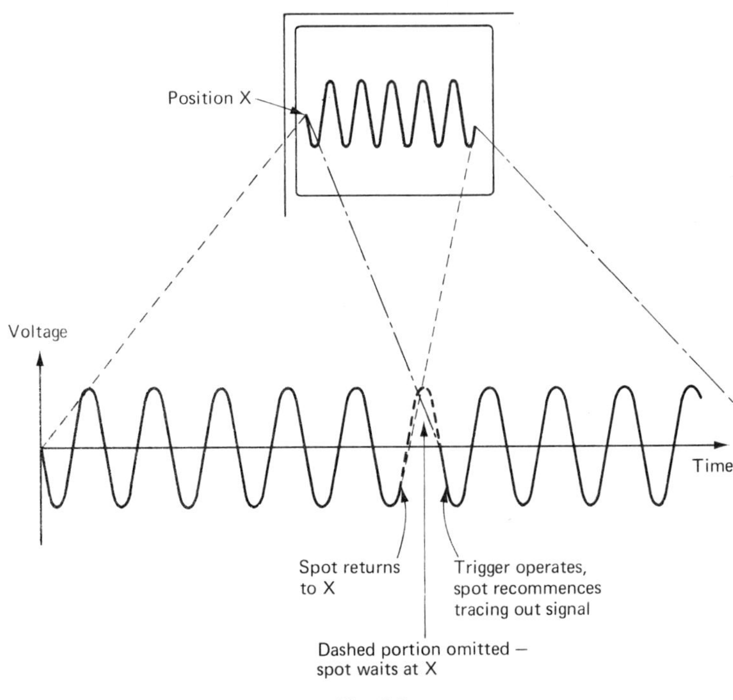


Fig. 2.5

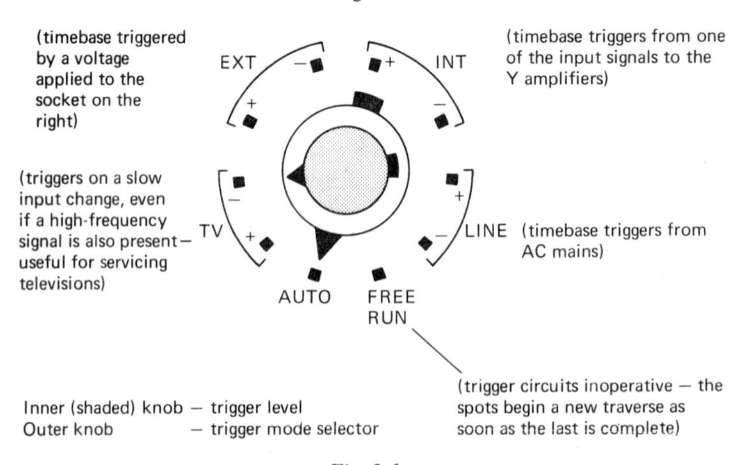


Fig. 2.6

The principle of triggering is that the timebase always begins to move the spot away from the left-hand end of the screen at some pre-arranged point in the cyclical variation of the input signal. The point chosen in the previous illustrations is the moment at which the input signal passes through its average central position in a negative-going direction (figure 2.5). The trigger circuits can operate on either of the two oscilloscope beams. The switch labelled 'TRIG SOURCE' in the centre of the left-hand panel enables you to decide which. The trigger circuit controls are at the bottom of the right-hand panel (figure 2.6). 'Internal +' and 'internal −' are the most commonly used settings. INT + triggers the timebase when the input signal selected is increasing in voltage (trace e, figure 2.7) and INT − when it is decreasing in voltage (trace c). The trigger-level control sets the voltage level at which the trigger

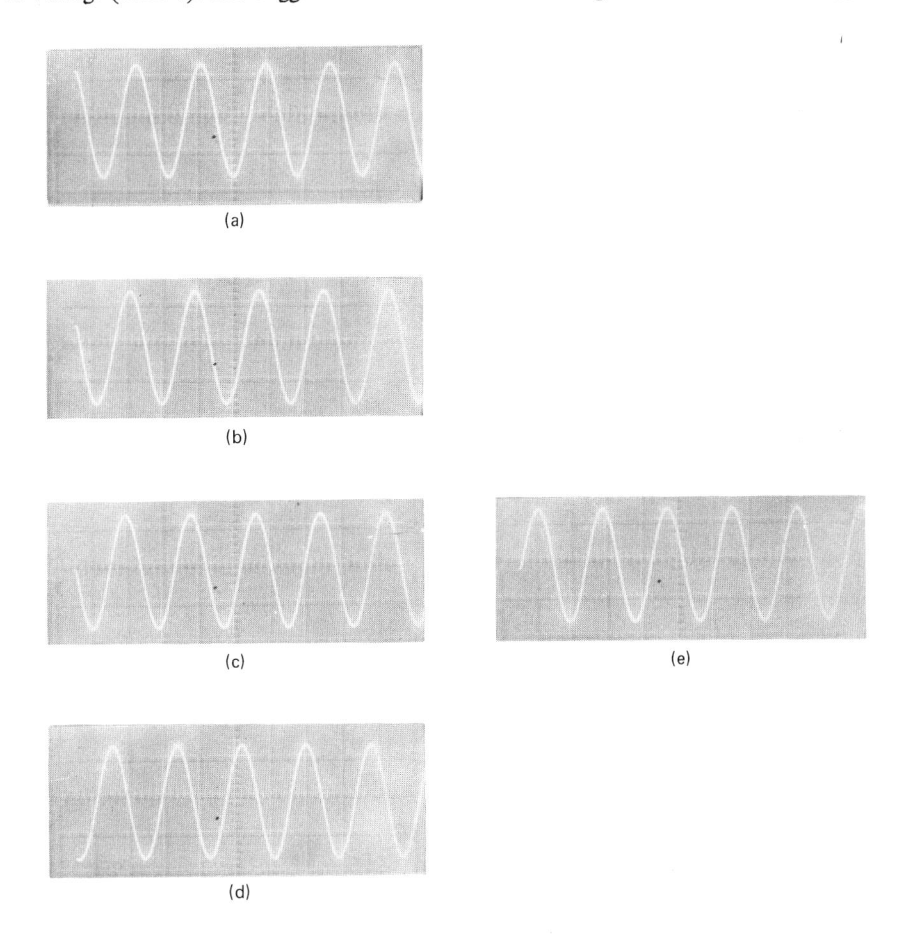


a, b, c and d have TRIG SELECTOR SWITCH in INT− position
e has TRIG SELECTOR SWITCH in INT+ position

Fig. 2.7

operates. Traces a, b, c and d in figure 2.7 show the effects of different settings of this control with the same input signal.

It is often possible to set trigger levels which the signal never reaches, in which case the timebase never triggers and the 'no picture' situation results. To prevent this from happening during normal use of the oscilloscope an automatic trigger-level control is provided as an alternative to selecting an appropriate one with the knob. To actuate this circuit you turn the trigger-level knob counterclockwise till a switch clicks, bringing in the auto-circuit. What this does is to set the trigger level at the average voltage level of the input signal. For example in the case of a sine wave this comes halfway up the display. Clearly on this setting the tiniest input variation will cause the oscilloscope to trigger, and in general, you get a bright horizontal line even in the absence of any input signal.

For the vast majority of oscilloscope uses leave the trigger-level control set to AUTO and the trigger-selector switch set to INT + or INT −; it does not matter which. The kind of situation in which the setting of the trigger-level control really matters is, for instance, when you are attempting to look at the shape of an individual 'spike' in a continuing but irregular chain of nerve impulses (figure 2.8).

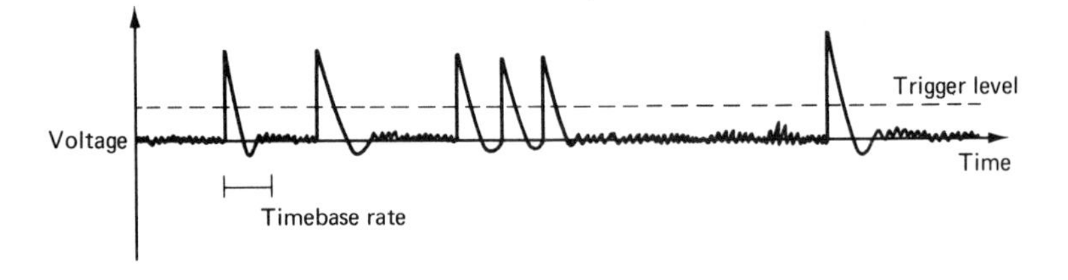


Fig. 2.8

On the AUTO setting the oscilloscope would trigger on the noise (irrelevant, unwanted, random signal) between the spikes. So the level control is used, and set so that the trigger operates when a positive-going signal passes through the dotted line shown, representing the voltage level selected with the control. The timebase speed is selected so that the traverse of the screen is completed in a time just longer than each spike. The timebase now triggers only when a spike occurs, and the resulting display shows the precise shape of the spike (figure 2.9).

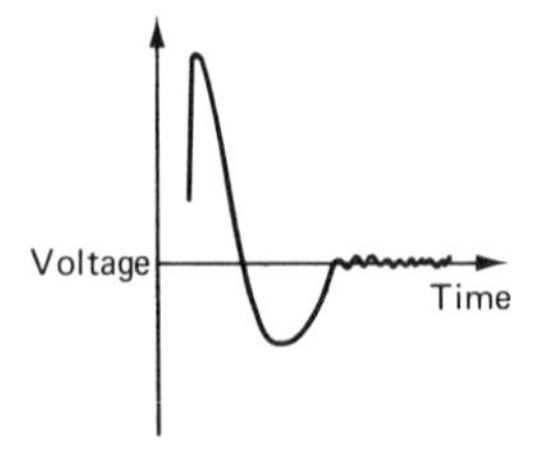


Fig. 2.9

Some oscilloscopes have an extra trigger-circuit control, a knob labelled STABILITY. If this is turned too far clockwise the display is unstable and moves back and forth; if it is turned too far anticlockwise the picture disappears altogether. Hopefully you can find an inbetween region of satisfactory triggering which gives a stable display.

## Single Shot Operation

Some oscilloscopes, not including the one in figure 2.1, have a facility called 'single shot' in which the timebase does a single sweep across the display tube each time a button is pressed, but otherwise remains inoperative.

This is useful for taking photographs of traces resulting from irregularly changing voltages, successive samples of which will not lie on top of each other to produce a steady picture. First move the switch from NORMAL to SINGLE SHOT (the picture disappears), open the camera shutter, then press the single-shot button (the spot appears at the left of the display, and the timebase moves it once across to the right), then close the camera shutter.

If your oscilloscope lacks a single-shot button, a very convenient way of achieving the same result is to use the camera flash contacts to connect a battery to the 'external trigger' socket (figure 2.10). To take a photograph you simply move the

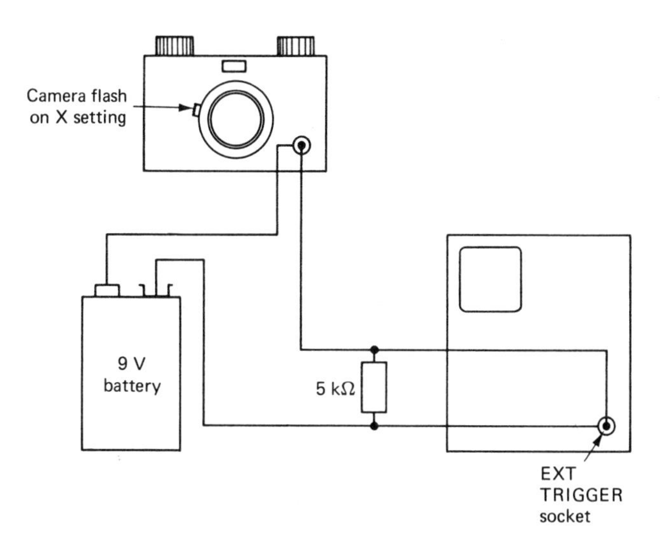


Fig. 2.10

trigger-selector switch to EXT and press the camera shutter release, having selected an exposure time a little longer than the spot traverse time.

It is worth checking that the single shot switch is in the NORMAL position when an oscilloscope is first switched on, otherwise no picture will appear.

## 2.3 AC-COUPLING – THE CAPACITOR

I have left till last one of the oscilloscope *Y*-amplifier controls, the AC/DC switch at the extreme left of the left-hand panel. This is a control which is quite subtle in its effects on the oscilloscopes behaviour, but very simple indeed in electronic terms

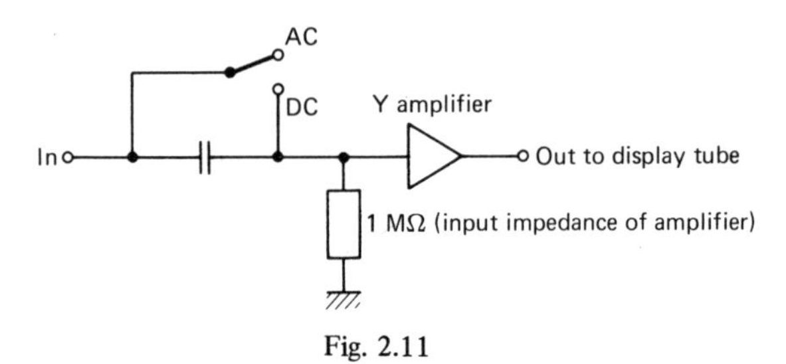


Fig. 2.11

(figure 2.11). It connects a component called a capacitor (⊣⊢) between the input socket and the input of the *Y*-amplifier in the AC condition, and shorts it out in the DC condition.

### Construction

A capacitor often consists of two long narrow strips of aluminium foil, sandwiched with thin plastic sheets so that they do not touch at all, rolled into a cylinder and then set in a plastic block (figure 2.12).

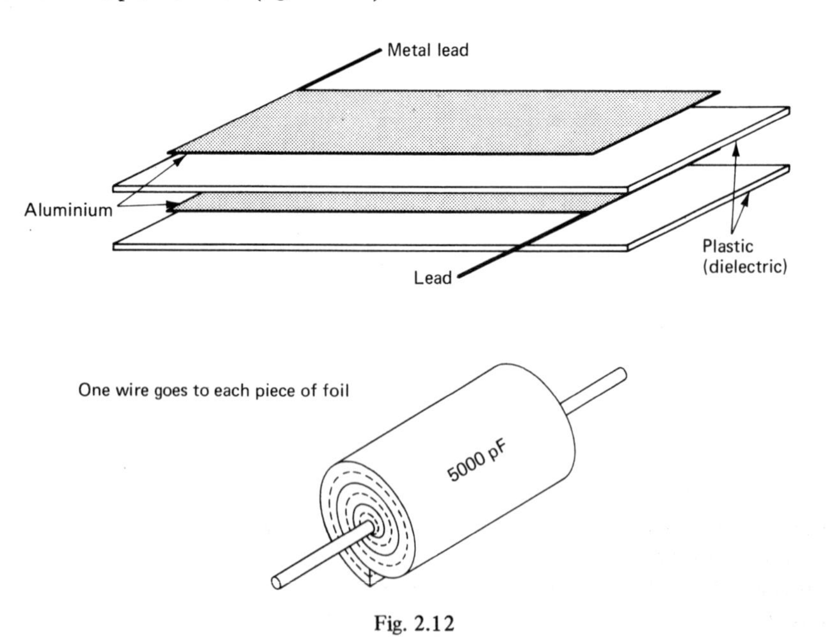


Fig. 2.12

**Electrical behaviour with changing (AC) voltages**

This component will not pass a steady voltage at all; there is, after all, a layer of plastic sheet in the way. Nevertheless it will transmit a changing voltage such as one of our sine-wave signals. The electrons moving into and out of the input metal strip as the signal voltage changes attract and repel electrons in the output strip, which moves in sympathy.

The faster the oscillation of a signal the more readily it is transmitted through a capacitor; and the larger the capacitor the lower the frequency of signal it is prepared to transmit at all. The size of a capacitor (proportional to the area of its plates, their closeness together, and the properties of the 'dielectric' or insulating layer) is measured in farads. This is an enormous unit, and microfarads ($10^{-6}$ farads, millionths of a farad), nanofarads ($10^{-9}$ farads, thousand millionths) and picofarads ($10^{-12}$ farads, million millionths) are commonly used. Nanofarads have only recently been introduced, and you still see lots of capacitors marked 50 000 pF or 0.05 $\mu$F when they mean 50 nF.

Clearly the electrical resistance of a capacitor to AC signals, called its *impedance*, varies with the frequency of the signals used. The formula for this is

$$Z = \frac{1}{2\pi FC}$$

where $Z$ is the impedance, $F$ is the frequency of the signal being used (in cycles per second, or Hz), and $C$ is the capacitance in farads. The theory behind this piece of electronics always measures frequencies in angles in radians, and hence $\pi$s keep turning up.

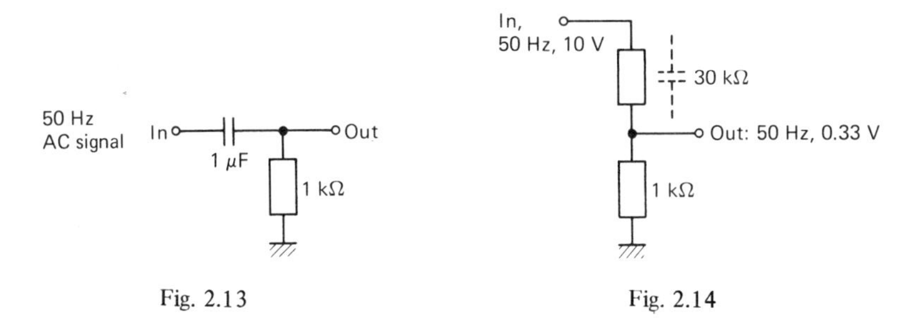


Fig. 2.13

Fig. 2.14

Taking the circuit shown in figure 2.13 we have

$$Z = \frac{1}{6\text{ (approx. }2\pi) \times 50\text{ (Hz)} \times 10^{-6}\text{ (farads)}}$$
$$= \frac{10^6}{300}$$
$$= 30\,000\ \Omega \text{ or } 30\text{ k}\Omega$$

In effect the situation is as shown in figure 2.14.

At frequencies so low that a given capacitor is heavily attentuating the input signal, there is a marked tendency for the metal plate on the output side of the component to respond to the rate of change of the input signal rather than the input voltage iself (figure 2.15).

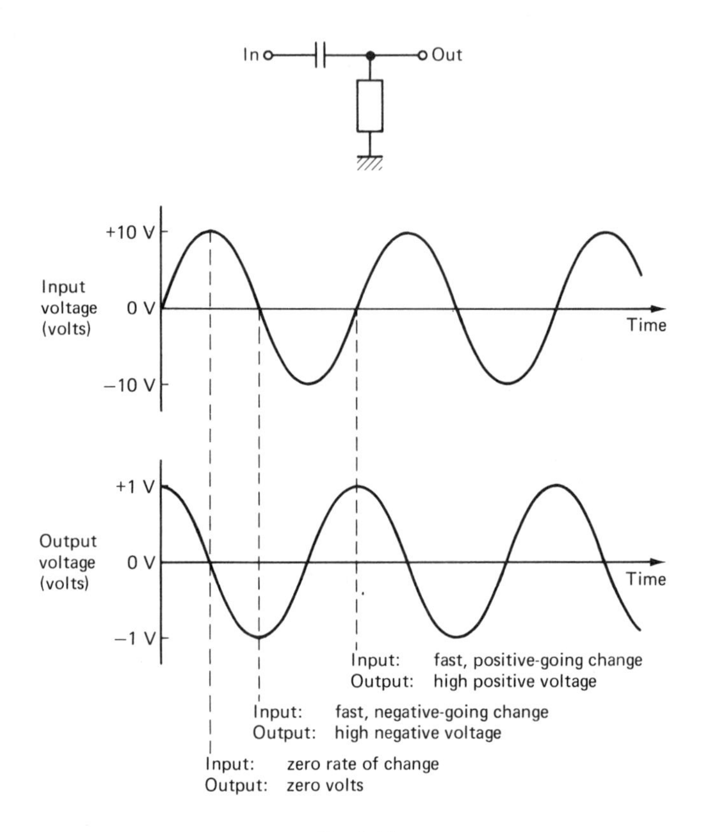


Fig. 2.15

**Electrical Behaviour with Steady Voltages**

Capacitors store energy when a voltage is applied across them (figure 2.16). When the switch is closed, point X goes up to +9 volts; when the switch is opened again point X stays at +9 volts, and the capacitor is said to be charged. Any attempt to measure the voltage at X with an oscilloscope, after the battery has been disconnected from the capacitor, quickly leads to the discharge of the capacitor (figure 2.17). The energy leaks away in the form of a current which flows to ground through the 1 MΩ oscilloscope input resistance. The larger the capacitor the longer it takes to discharge.

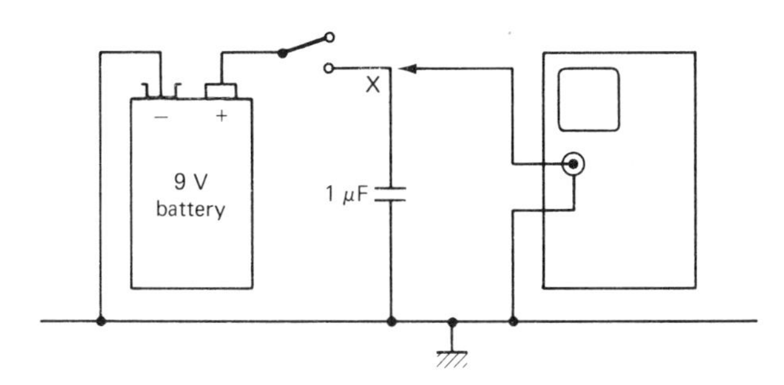


Fig. 2.16

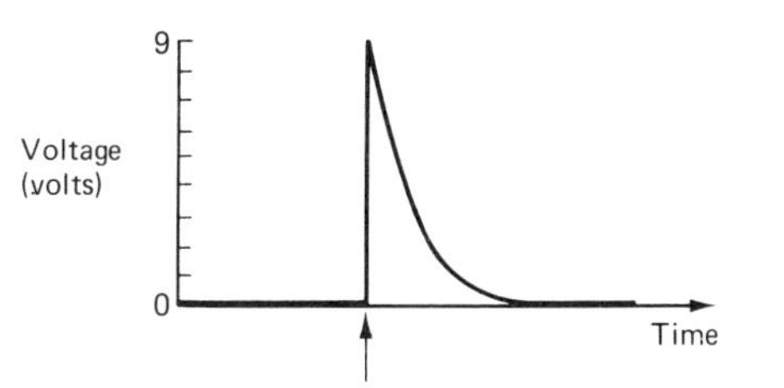


Fig. 2.17

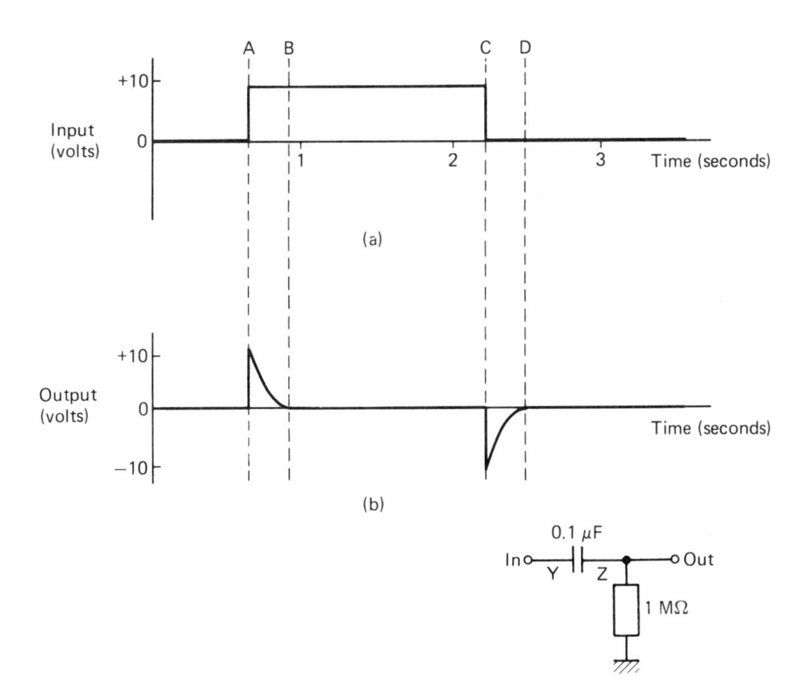


Fig. 2.18

**AC-coupling**

AC-coupling two circuits means joining them together via a capacitor. If we apply a signal (figure 2.18(a)) to an oscilloscope with its input switch in the AC position (the actual input circuit is shown below), having selected a very slow timebase speed, the resulting trace takes the form shown in figure 2.18(b). We have two ways of interpreting this. First, in behavioural terms, the output of an AC-coupled circuit tends to give the rate of change of the input signal. At time A there is a high positive rate of change on the input side, so the output is a positive voltage. At time B there is no input change, so the output is zero volts. At time C the input voltage moves sharply negative, so a large negative output results.

Second, viewing the same series of events in electronic terms, the initial fast voltage change is transmitted through the capacitor giving the positive spike at time A. The capacitor then begins to charge and point Z on the circuit returns to zero volts as the capacitor becomes fully charged. It remains thus until time C, when the input signal carries point Y down to zero volts. This sudden change gives the capacitor no time to discharge, so immediately after time C there is still 10 volts across the capacitor, and hence Z has to be at – 10 volts. The capacitor now discharges by passing current through the 1 MΩ oscilloscope input resistance until Z reaches zero volts at time D.

The AC-coupling switch on an oscilloscope is most frequently used in the situation where the signal is a small oscillation (AC component) superimposed on a large steady voltage (DC component) (figure 2.19(a)). If this signal were applied to a DC-coupled input and the *Y*-scaling switch set to a sensitive enough level to allow the AC component to be examined carefully, then the DC component would carry the

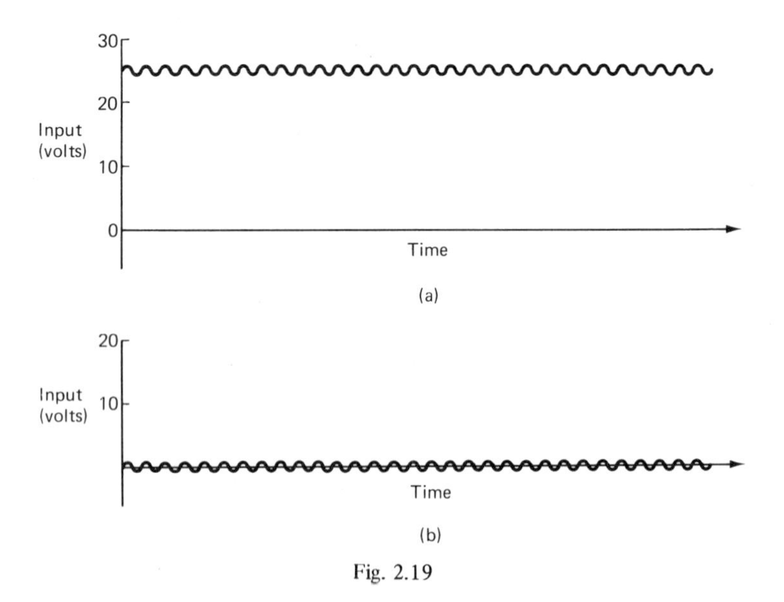


Fig. 2.19

picture so far off the top of the screen that the shift control would be unable to bring it back again. The AC-coupling circuit removes the DC component and the signal in figure 2.19(b) forms the actual amplifier input, thus removing the problem.

One common snag with AC-coupled inputs is illustrated by the situation on p. 21, which involves a battery used to trigger an oscilloscope via the external trigger socket. This is generally AC-coupled to the trigger circuit (figure 2.20). If you leave

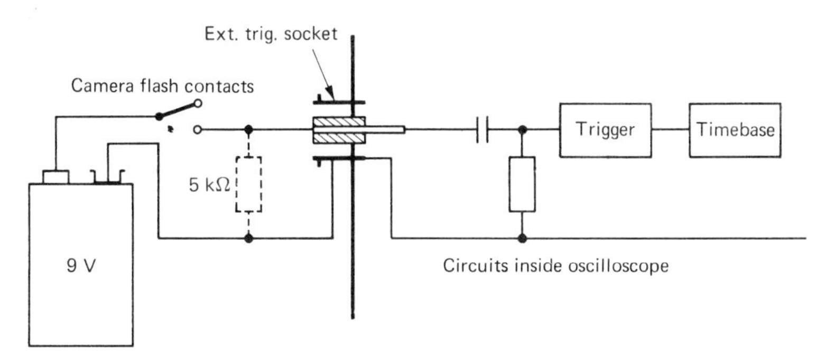


Fig. 2.20

out the dotted resistor the oscilloscope triggers the first time you push the switch, but fails to do so for subsequent pushes. If you leave it alone for half an hour it will respond again, but only once. The cause of this problem is that, on the first push, the capacitor transmits the sharp voltage rise from the battery when the switch is closed, then charges up fully. When the switch is released, the situation is as shown in figure 2.21 and the capacitor has no way of discharging except by internal leaks,

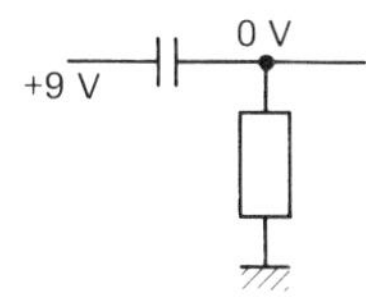


Fig. 2.21

which take a very long time. So if the switch is closed again nothing further happens. The solution is to supply the dotted resistor, through which the capacitor can discharge itself.

## 2.4 OSCILLOSCOPE DISPLAY SYSTEMS – THE STORAGE OSCILLOSCOPE

The display on a normal oscilloscope tube (a cathode ray tube) is produced by a phosphor coating on the inside of the vacuum tube which glows when an electron beam falls on it. The electrons are produced by a heated wire filament (the cathode) which throws off electrons into the vacuum, and a beam is formed by attracting these negatively charged particles to a positively charged anode (figure 2.22). The anode has a small hole which the beam shoots through and then passes between two sets of metal plates, arranged at right-angles to each other (figure 2.23). When voltages are applied across these plates, the beam is deflected by an amount

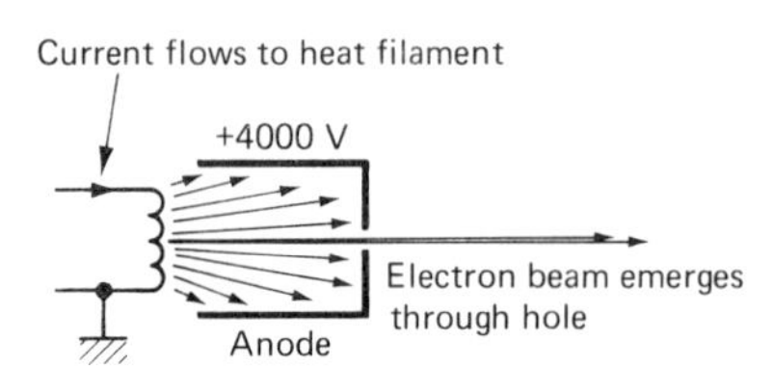


Fig. 2.22

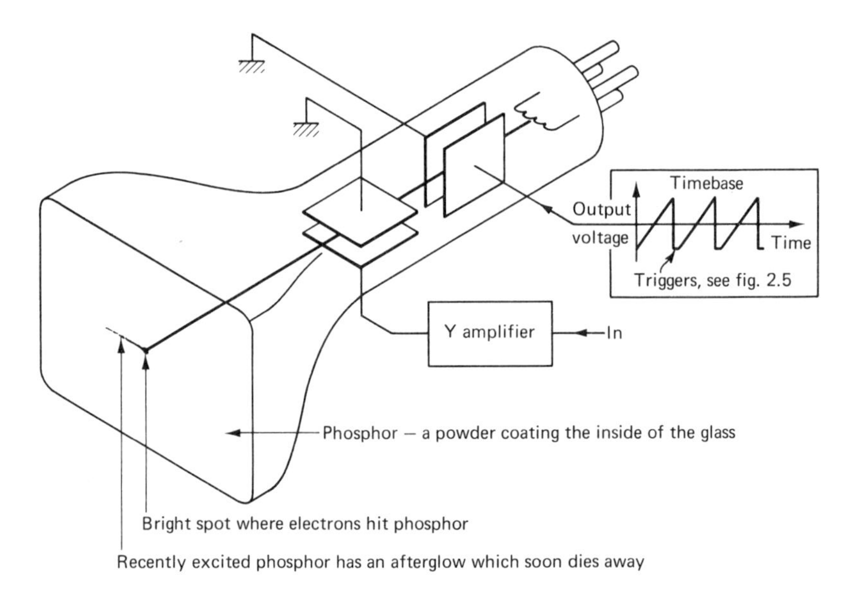


Fig. 2.23

proportional to the applied voltage. The set of plates which deflect the beam horizontally are connected to the timebase and those which deflect it vertically to the *Y*-amplifier.

This standard oscilloscope tube is very satisfactory for studying signals which repeat themselves and produce a steady display when the trigger operates properly, but it is not much good for looking at the isolated events which so frequently interest us in biology. Even with a long-persistence phosphor, which goes on glowing for 1 or 2 seconds after the beam has excited it, the display is too labile for the observer to extract useful information. Attempts to photograph traces from such events often fail because you tend to press the camera shutter button a fraction of a second too late. The solution is obviously a display tube which stays illuminated indefinitely once an electron beam has passed over it. This is achieved by the storage oscilloscope, which in addition to its normal functioning as an ordinary oscilloscope can also store the display as a pattern of electric charges on a grid just behind the phosphor.

A sample of the signal is stored by switching the timebase to single shot operation, pressing the store button, then releasing a single timebase sweep. As the spot passes across the tube face it leaves a brightly lit trace behind it. This trace can be closely examined, photographed, or even copied with tracing paper. Pushing the erase button erases it, leaving the screen free to store another sample. Releasing the store button returns the oscilloscope to conventional operation. Two buttons are provided, one for the top half of the display screen and one for the bottom half, so it is possible to work with one beam storing and the other operating conventionally.

To store a rare event it is best to leave the timebase running, push the store button and turn the brightness down to a level at which the beam will only just write on the storage grid. The event will be seen superimposed on a horizontal line stored while nothing was happening. It is also possible to see the average pattern of an event by storing many repetitions on top of each other.

# 3 Power Supplies

## 3.1 BATTERIES

A lot of equipment needs a DC low voltage supply, such as you get from batteries or accumulators, and there is a natural temptation to use these rather than building or buying power supplies which work from the AC mains.

This often leads to disaster, because the batteries frequently go flat when you are just about to make a crucial discovery. If batteries are inevitable, choose big ones, six 1.5 volt torch cells being much more satisfactory than a 9 volt transistor radio battery. Also replace them regularly, say on the first of every month whether the equipment has been used or not.

## 3.2 AC MAINS

The earth wire is generally connected to a water pipe, and represents our standard zero ground potential. The blue and brown wires are ultimately connected to the two terminals of the generator (figure 3.1).

All the current taken from the brown wire must flow back to the generator via the blue one, which is earthed at the sub-station and so stays very close to zero volts. But it is very unwise to rely on the blue wire being nearly at earth and therefore safe, for it only takes a reversed set of connections somewhere in the system to swap over the live and neutral wires. No piece of equipment should have the mains neutral connected to its chassis, or should involve any connection of neutral to earth.

## 3.3 TRANSFORMERS

### Mains Transformers

A transformer consists of two coils of insulated wire wound round an iron core (figure 3.2). When a current flows through the primary coil or 'winding' the iron

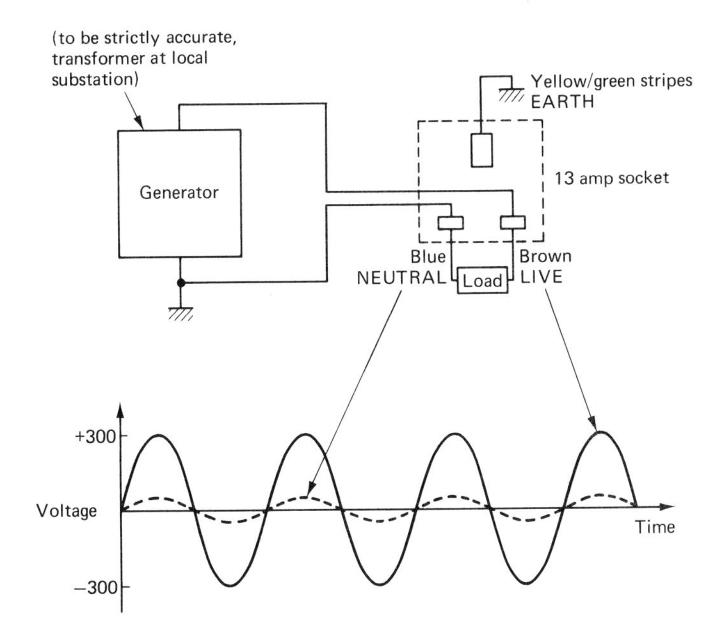


Fig. 3.1

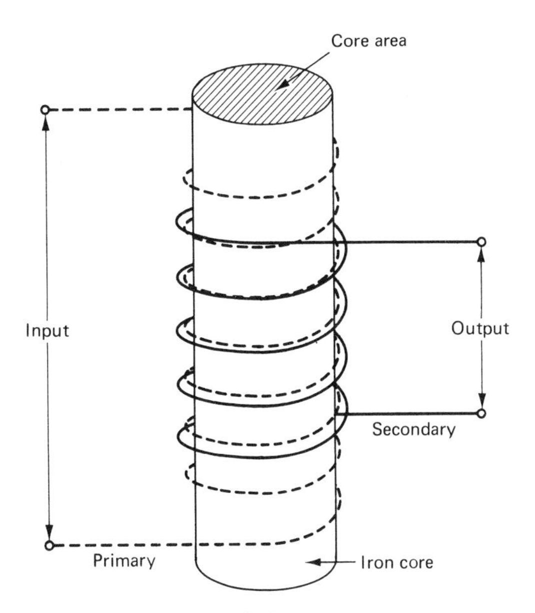


Note both coils have same sense
(clockwise here)

Fig. 3.2

bar becomes an electromagnet; which end is a north pole depends on the direction of the current flow. So if an AC mains current flows through the primary, the core will be a magnet whose north pole changes from end to end 100 times a second. From the point of view of the secondary coil, it is as if it were rotating end over end in a magnetic field, exactly as the coil of a dynamo does. It is hence possible to draw a current from the secondary coil. The AC voltage across the secondary coil depends

(1) on the AC voltage across the primary coil; halving the primary voltage halves the secondary voltage.

(2) on the number of turns in the two coils.

$$\frac{\text{Volts in}}{\text{Volts out}} = \frac{\text{Number of turns on primary coil}}{\text{Number of turns on secondary winding}}$$

So a given transformer can be used either to increase or decrease voltages depending on which coil is used as the primary (figure 3.3). (However, remember that the

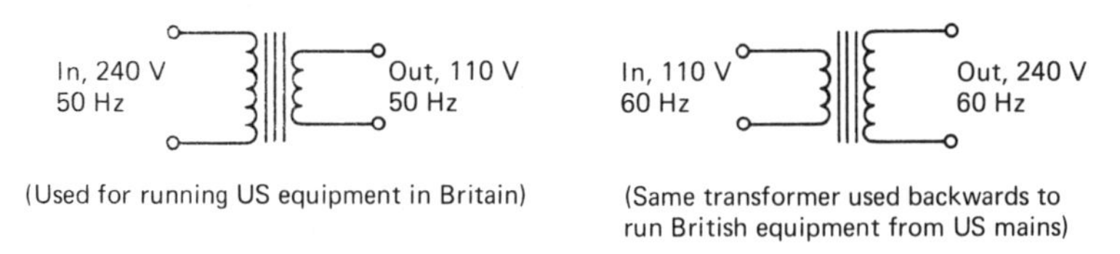


Fig. 3.3

speed of many clock and tape recorder motors depends on the *frequency* of the supply.) This reversibility depends on the insulation to some extent; for instance if you tried to use a transformer designed to produce 12 volts from 240 volt mains to produce 4800 volts from the mains by connecting it backwards, it would almost certainly destroy the insulation and burn out.

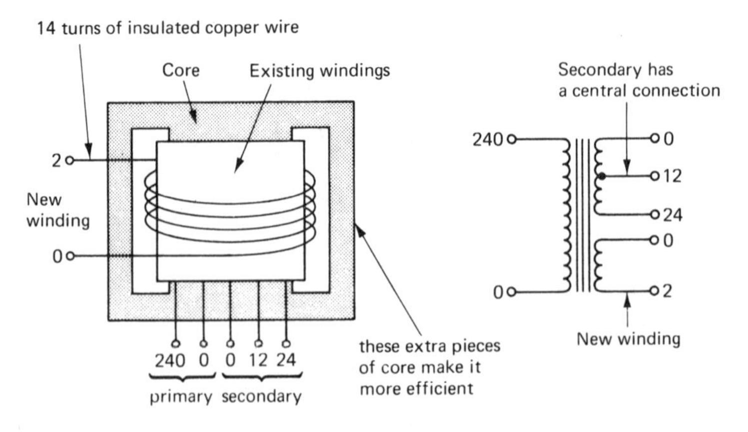


Fig. 3.4

Transformers are designed with a specific frequency of operation in mind. A mains transformer could not be used for frequencies higher than, say 150 Hz, and a transformer designed to work in a VHF radio would greatly attenuate low frequencies.

The ratio of the number of turns of the two coils sets the input/output ratio of a transformer, but the absolute number of turns used depends on the current requirements and the sort of core used. A rough guide for mains transformers is that there are generally 7 turns per volt for a core area of one square inch or less. Bigger cores have 6 turns/volt. This may be useful if you are trying to add an extra secondary coil, for example to run a 2 volt lamp, to an existing mains transformer (figure 3.4).

There is a limit to the current you can draw from a transformer because there is a limit to how strong an electromagnet you can make a given piece of iron, however much current you pass through the coil surrounding it. The smaller the core, the lower the current at which this saturation effect occurs.

As there is no direct electrical connection between the two coils the transformer transmits no steady (DC) element of the input, and it is possible to connect either end of the secondary winding to any voltage you like (figure 3.5). The output of a transformer is said to float.

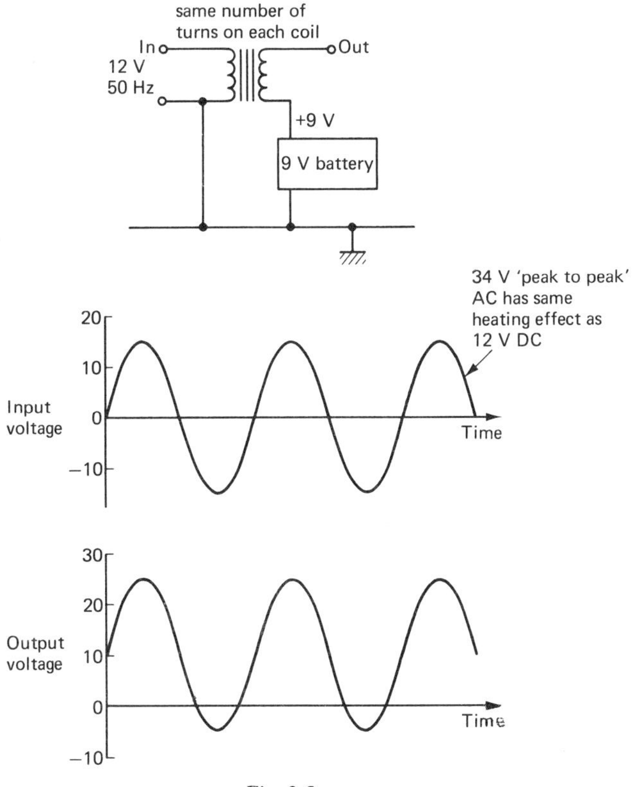


Fig. 3.5

**Autotransformers**

Autotransformers have only one coil, which serves as both a primary and a secondary. The output currents are those induced by the changing magnetic field, which is itself produced by the current already flowing in the coil. Autotransformers are often variable (figure 3.6). This design saves on copper wire, but does have an

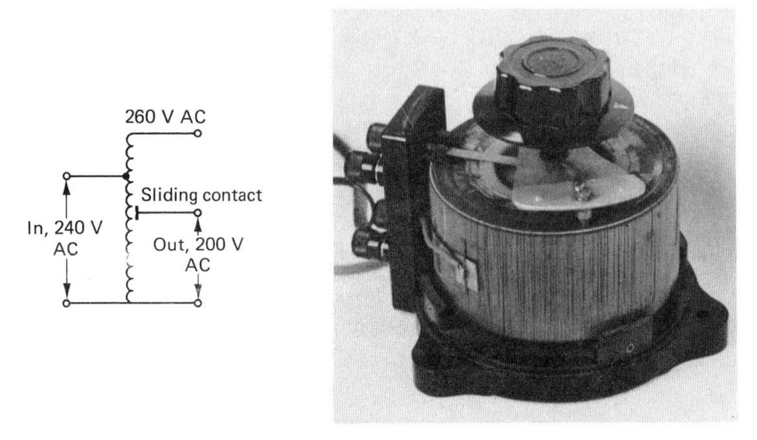


Fig. 3.6

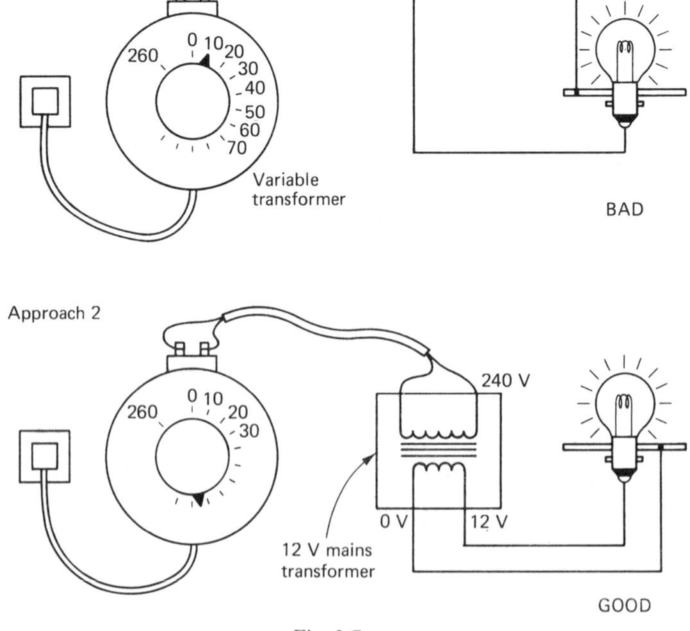


Fig. 3.7

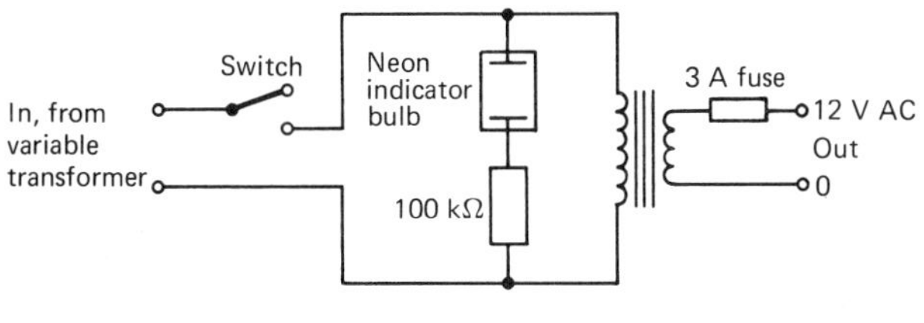


Fig. 3.8

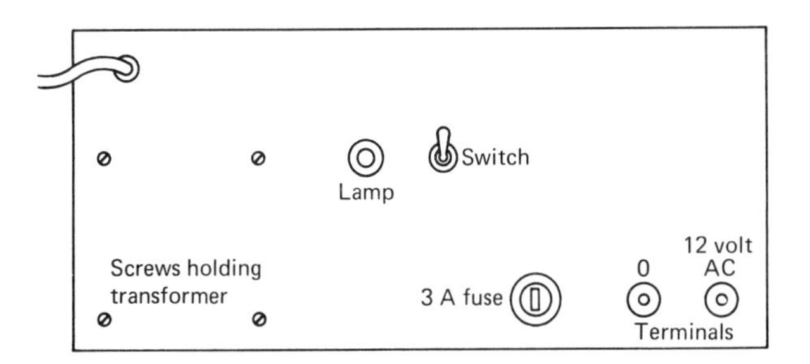


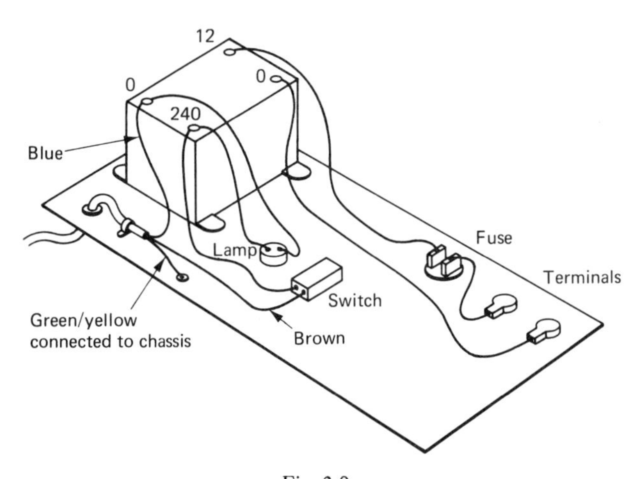


Fig. 3.9

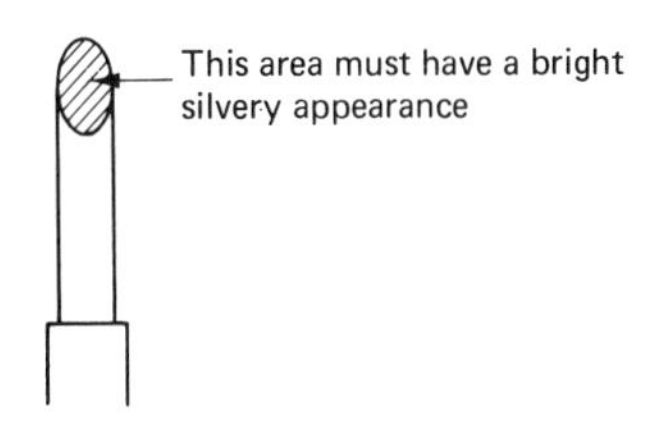


Fig. 3.10

output connected at one end to the blue neutral mains wire. Variable transformers are useful for dimming lamps, running motors slowly, etc., but their use calls for care. For instance if the problem is to drive a 12 volt car headlamp bulb, one terminal of which is connected to the chassis of the apparatus (figure 3.7), approach (1) is bad because (a) if the mains plug is wired the wrong way round, the chassis will be live; (b) one slip, and you turn the voltage up to 260 volts and blow the bulb. In approach (2) the transformer means that the bulb supply is isolated from the mains, and moreover the whole scale of the variable transformer is useable.

## Constructional Details

The 12 volt transformer should be mounted in a metal box, preferably with a fuse and switch, and a neon indicator (figure 3.8). Everything should be attached to the front panel, so that the rest of the box lifts completely clear for easy assembly and service (figure 3.9). The 100 kΩ resistor prevents the neon lamp from drawing too much current; sometimes the resistor will be built into the lamp assembly already. Always use standard plastic-covered copper wire for connections. To make the joints you need a *small* (15 watt) electric soldering iron and some thin flux-cored solder. The iron end must be shiny (figure 3.10); it is no use soldering with something that looks like a well-used poker. If you have an iron in this state attack it with a file till shiny copper shows, switch on, and, as soon as it is hot, touch it with the solder so that it gets a shiny coat of solder. This process is called tinning. To join a wire onto a tag, you first tin both the wire and the tag, then lay the tinned wire on the tinned tag, apply the iron to the wire till all the solder melts and joins up, remove the iron and hold the wire still briefly until the solder sets (figure 3.11). If

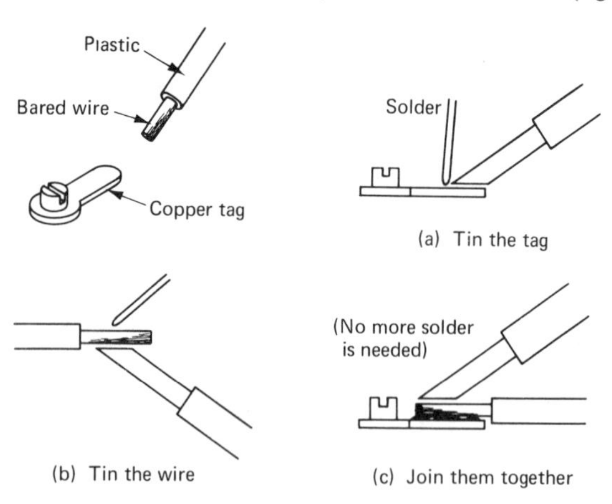


Fig. 3.11

the joint looks bright and shiny it is all right; if it looks crystalline and powdery it is not. Such 'dry joints' lead to endless trouble because they are mechanically weak and conduct poorly and erratically.

## 3.4 DIODES

Diodes are components whose resistance changes in a way which depends on the sign of the voltage applied to them. Figure 3.12 shows the symbol for a diode and illustrates the manner in which it works. Provided that there is a positive current

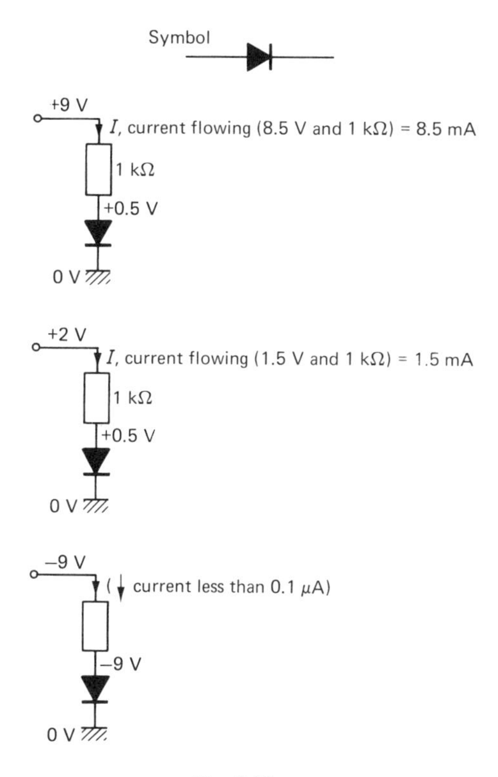


Fig. 3.12

trying to flow through the diode in the direction of the arrow (of the symbol) the device has a low resistance which depends on the current's size in such a way that the voltage across the diode is always about half a volt. If the positive current is lower than about 1 mA the diode's resistance rises dramatically, and similarly for negative-going currents the diode's resistance becomes many megohms. So a diode works as a kind of switch, as shown in figure 3.13. Diodes can be used to convert AC mains into DC, a process known as rectification (figure 3.14). The DC you get with only one diode is very intermittent – this is called half-wave rectification – and a much more satisfactory state of affairs occurs with two diodes and two transformer secondary coils (figure 3.15).

Whichever end of a transformer coil you earth, the voltage difference across the

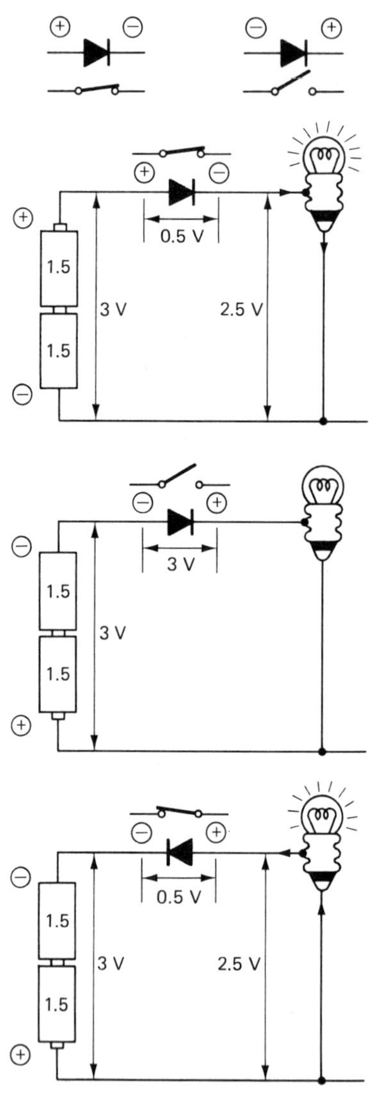
0.5 V
1.5
1.5
3 V
2.5 V
3 V
1.5
1.5
3 V
0.5 V
1.5
1.5
3 V
2.5 V

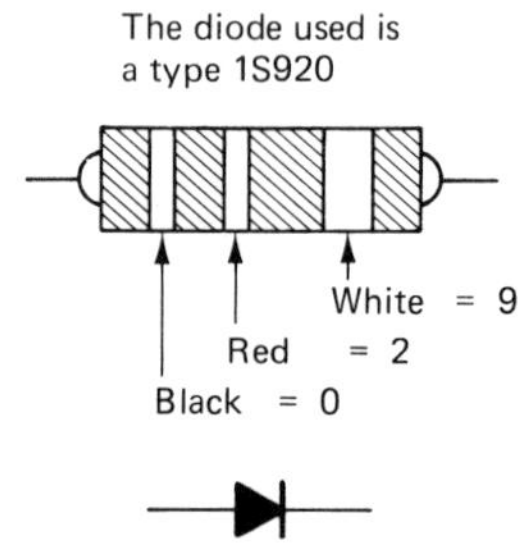
The diode used is
a type 1S920
White = 9
Red = 2
Black = 0
Broad white band
corresponds to bar
on the symbol
The number has no
electrical significance

Fig. 3.13

coil at any instant remains unchanged in size and direction. When the earth connection on the lower secondary coil is altered in circuit (2) graph (b) becomes graph (d). In every case the voltage difference between the two ends of the coil (vertical distance between full and dashed lines) is the same on the two graphs. The full line, the voltage at the top end of the coil, has been pushed from a sinusoid in graph (b)

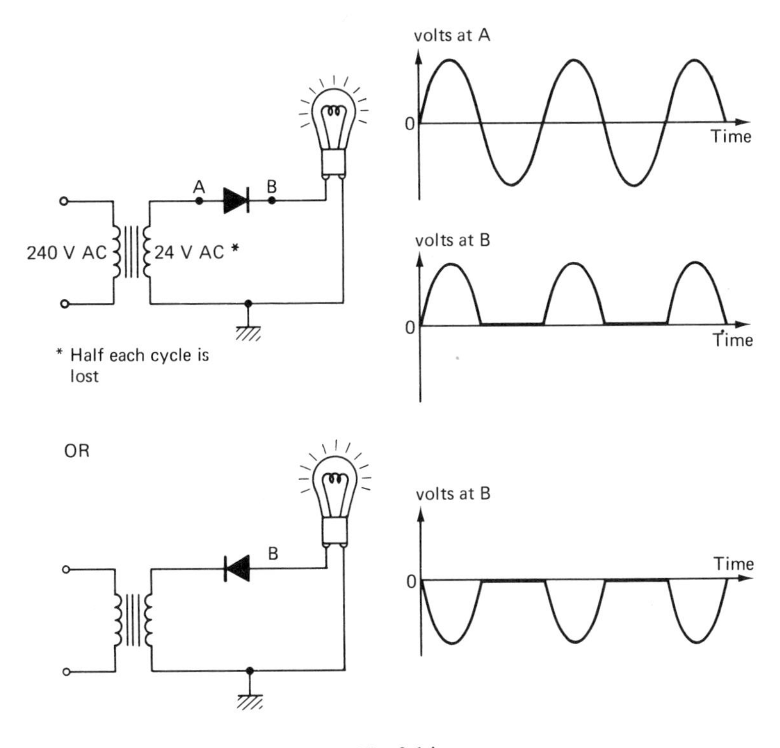


Fig. 3.14

to a straight line in graph (d) because the top end of the coil is now earthed. Meanwhile the dashed line, the voltage at the bottom end of the coil, is freed and is forced into an inverted sine curve to maintain the voltage differences across the coil. As shown in circuits (4) and (5) the two coils are often joined inside the transformer, and only a single external connection, the 'centre tap' of the two coils considered as one, is provided.

Full-wave rectification is possible with a single transformer winding. It needs four diodes and the rather confusing circuit shown in figure 3.16. Neither side of the transformer secondary winding is earthed except through the diode network, so all we need to consider is the voltage difference between A and B, which goes, in the course of one cycle of the mains, from being + 16 volts (A relative to B), through zero volts, to − 16 volts (A relative to B). When the difference is positive,

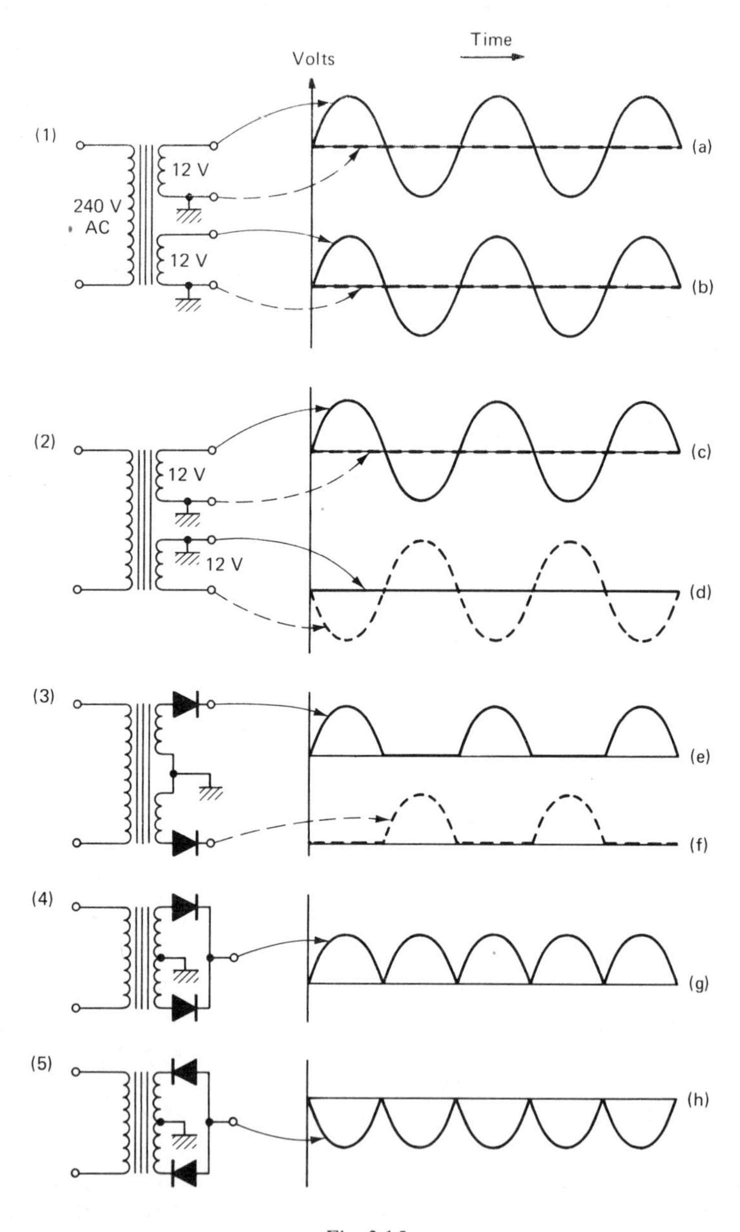


Fig. 3.15

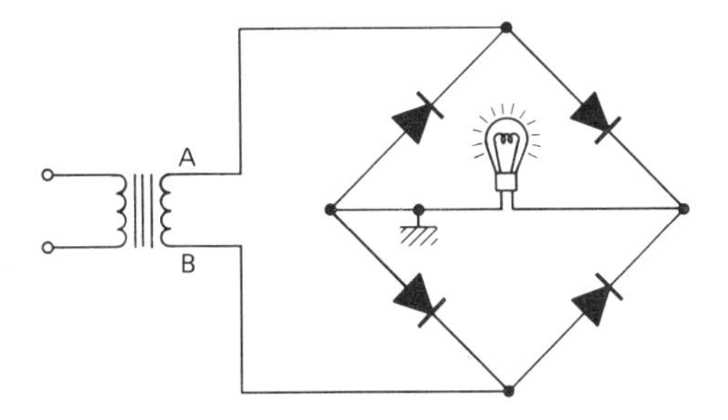


Fig. 3.16

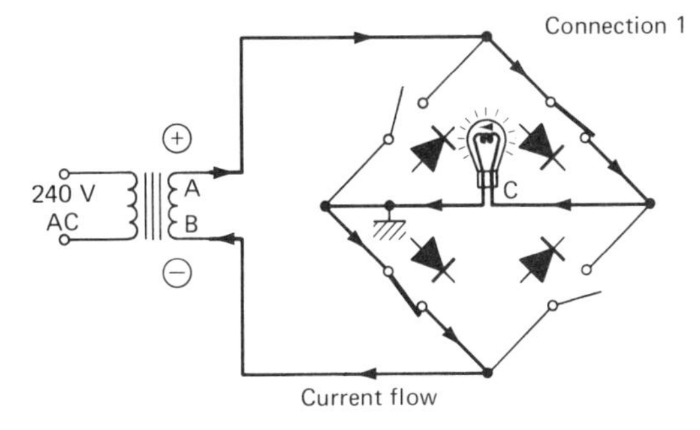


Fig. 3.17

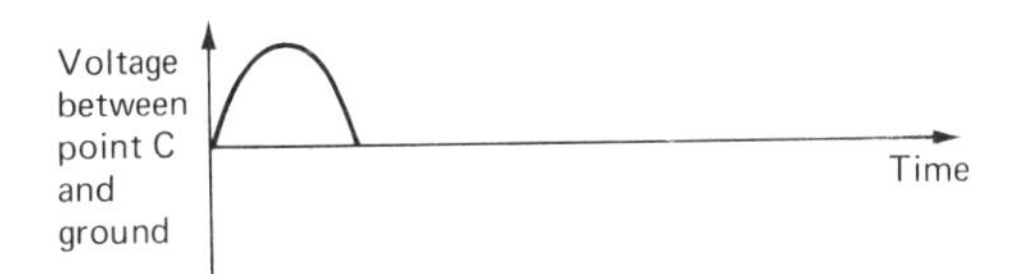


Fig. 3.18

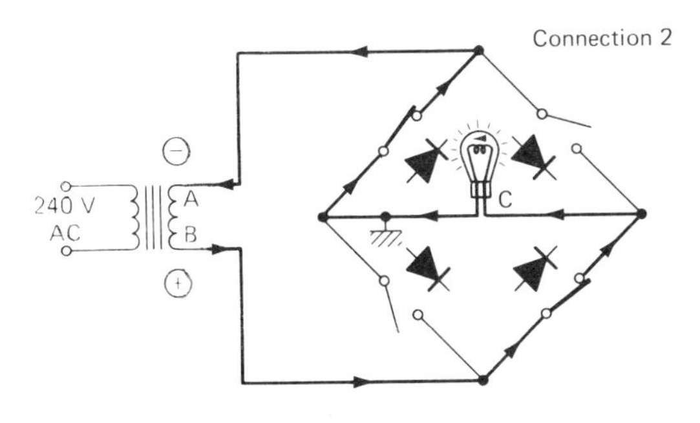


Fig. 3.19

two of the diodes are 'switched on' (figure 3.17). Point B is connected to ground, point A is connected to the lamp, and the situation is as shown in figure 3.18. Then, when the voltage difference between A and B becomes negative the diodes are forced to change roles (figure 3.19); point A is now connected to ground and point B to the lamp. The result of all this is shown in figure 3.20. The only earth

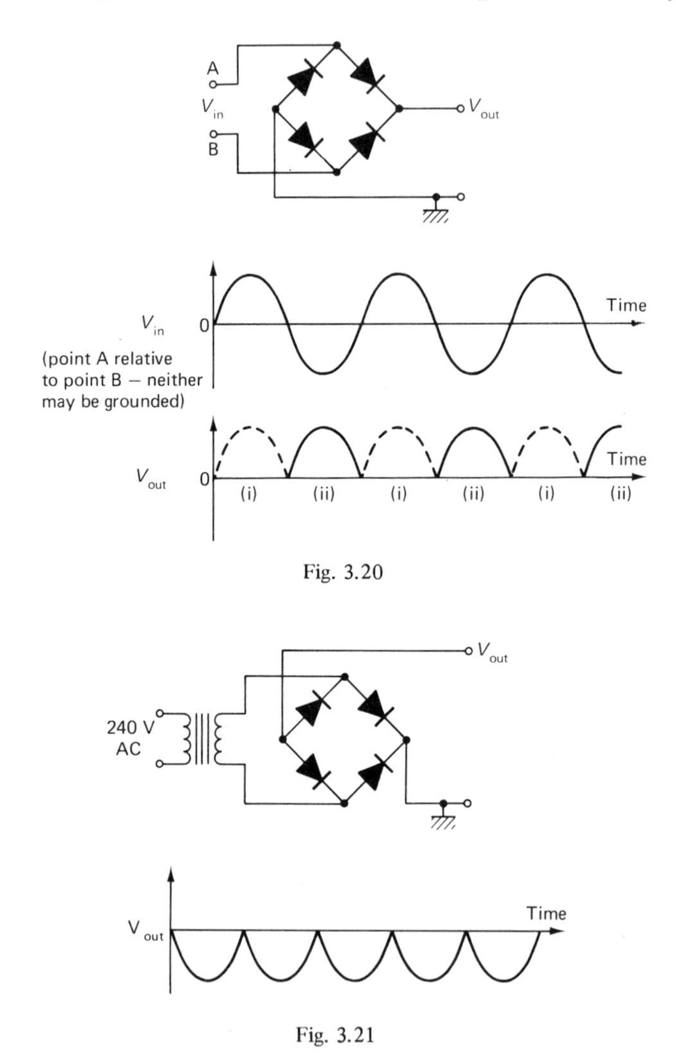


Fig. 3.20

Fig. 3.21

connection is on the output side of the bridge, and we can make the supply negative just by changing it over (figure 3.21). Getting diodes the wrong way round in building this circuit is common and leads to instant burning out of the diodes (they sometimes light up dull red) and sometimes even the transformer if you are slow about switching it off.

Fortunately manufacturers have succumbed to popular pressure and produced a full-wave bridge with all four diodes set in one plastic block. These have only four terminals, two labelled 'input' and two labelled 'output', and are foolproof.

Diodes come in a vast range of type numbers, but vary only in two important ways.

(a) The maximum voltage it is safe to have across the diode when the switch is open. This is called the peak inverse voltage (PIV), and you can buy diodes with PIV's ranging from 10–10 000 volts.

(b) The maximum current you can pass through the diode when the switch is closed. This is the average rectified forward current. Generally you are allowed to put ten times this current through the device for a brief part of each mains cycle, provided the current averages out over a few mains cycles.

## 3.5 PEAK AND AVERAGE VOLTAGES

If a 12 volt car headlamp bulb is connected to the secondary of a 'twelve volt' mains transformer it gives out exactly the same amount of light as it would if run from a source of 12 volts DC supply such as an accumulator.

This is because the peaks of the AC supply are extended to + 17 and − 17 volts to make good the power loss in the gaps (figure 3.22). AC peak voltages are always 41 per cent greater than the equivalent DC voltage. In fact the shaded area in figure 3.22 does not quite equal the hatched one, for power = current × voltage (see chapter 1) = voltage$^2$/(load resistance), so it is only for a graph of power against time that this equality holds (figure 3.23).

In practice a rectifier always diminishes voltages which pass through it because a diode in the conducting state ('switched on') always has 0.5 volt across it; so with the bridge circuit shown, in which there are always two switched on diodes connected in the circuit, we lose 1 volt, bringing out peaks down to 16 volts and our equivalent DC volts ('RMS' or root mean square) down to 11.3 volts.

It would seem at first sight from this that there is no hope of ending up with a steady 12 volts DC from a power supply whose transformer is rated at 12 volts. In fact it can be arranged that smoothing capacitors, dealt with in the next section, store a little extra power at the peak of each cycle to make good the power being wasted in the diodes, provided, that is, that the supply is not being expected to provide the full current rating of the transformer.

## 3.6 CAPACITATIVE SMOOTHING

### The Single Smoothing Capacitor

In Figure 3.24, the capacitor serves to convert the rectified AC of (a) into the DC with a slight AC ripple shown in (b). The capacitor stores power by charging up on

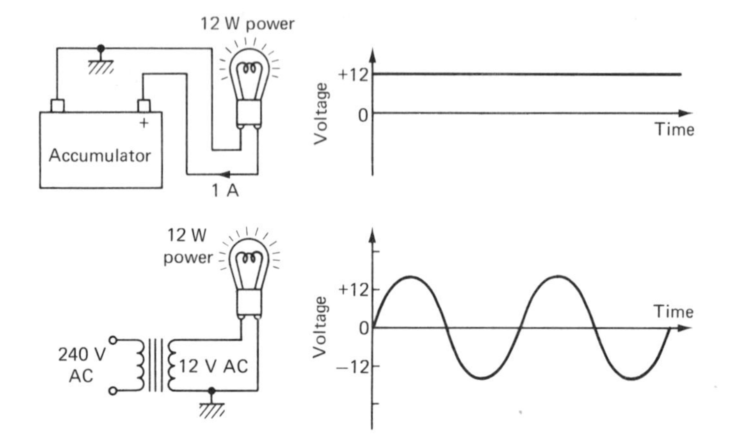


The way in which the extra voltage in the peaks in the AC supply 'fills in the gaps' is clearest after the AC has been passed through a full wave rectifier:

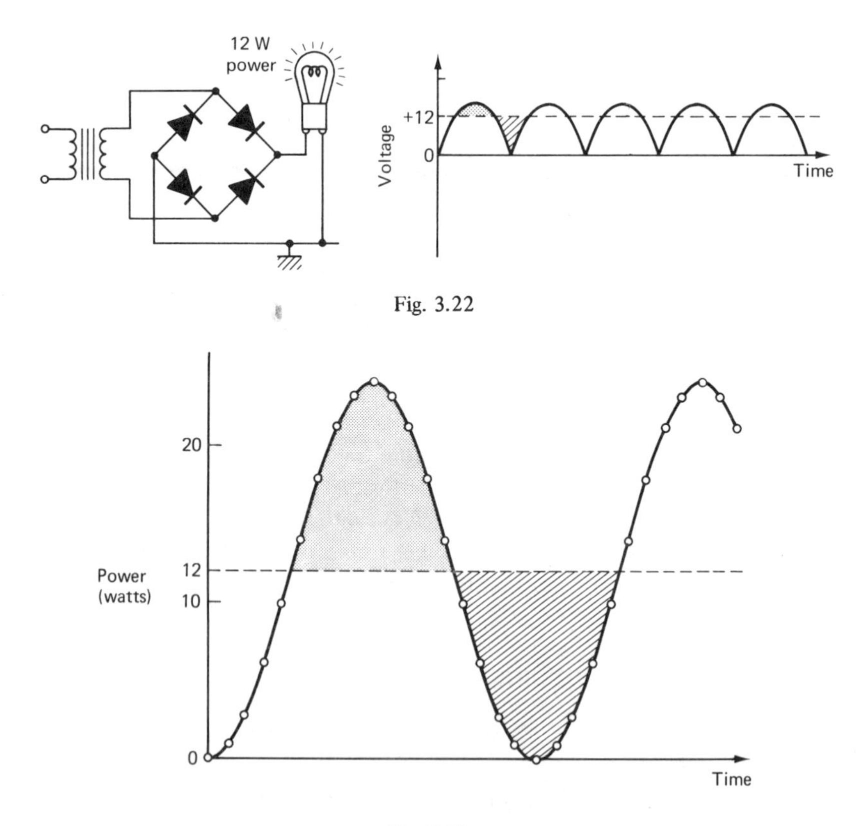


Fig. 3.22

Fig. 3.23

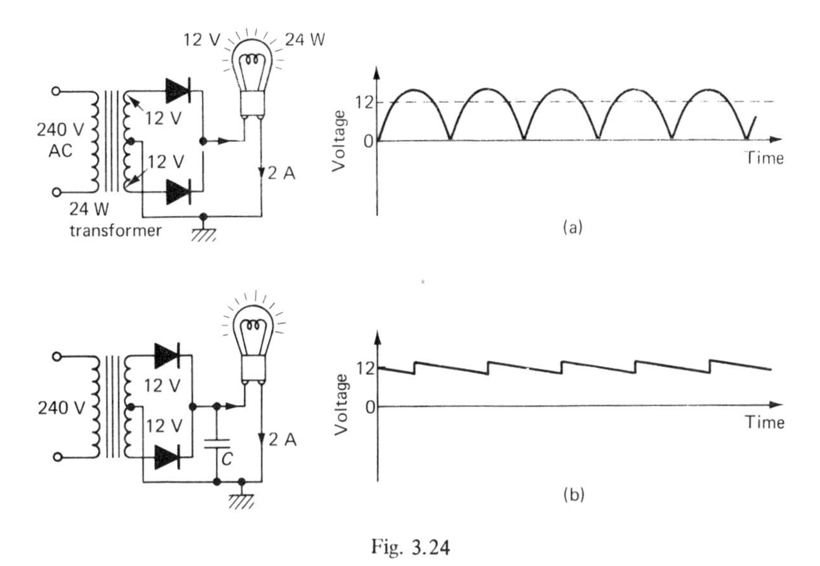


Fig. 3.24

the rectifier peaks, and then fills in the gaps by discharging through the lamp. Taking each gap to be 1/100 of a second we can use the basic equation for capacitor charge or discharge:

$$\frac{\text{change of voltage}}{\text{time taken}} = \frac{\text{current}}{\text{capacitance}}$$

$$\frac{\mathrm{d}V \text{ (volts)}}{\mathrm{d}t \text{ (seconds)}} = \frac{I \text{ (amps)}}{C \text{ (farads)}}$$

Fig. 3.25

($\mathrm{d}V/\mathrm{d}t$ is a mathematical expression meaning rate of change of voltage, in volts per second, at any given instant.) To predict how large a capacitor we will need to meet a given ripple specification, say 2 volts in the case shown in figure 3.25, we substitute values for the symbols in the above equation.

$$\frac{dV}{dt} = \frac{I}{C}$$

$$\frac{2\ \text{(volts)}}{\frac{1}{100}\ \text{(seconds)}} = \frac{2\text{(amps)}}{C\ \text{(farads)}}$$

$$\therefore C = \frac{1}{100}\ \text{farads} = 10\,000\ \mu\text{F}$$

Capacitors as big as this are invariably *electrolytic*. The large capacitance is obtained by having an extremely thin layer of dielectric formed chemically between the capacitor plates, one of which is a conducting paste. Electrolytic capacitors must be connected into the circuit the right way round: the end marked +, or made of plastic or rubber rather than metal, must go to the positive voltage side of the rectifier (figure 3.26). Failure to get this right results in the capacitor exploding and getting a smelly corrosive white chemical everywhere.

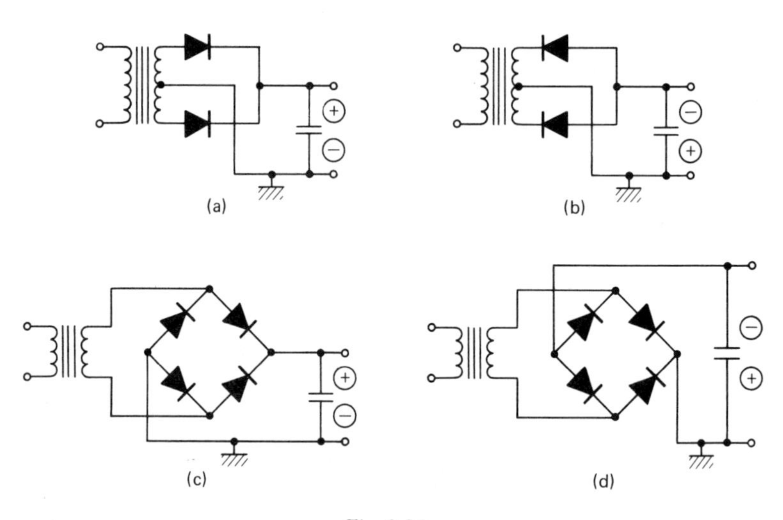


Fig. 3.26

It is also important to choose a capacitor of suitable *working voltage* (the maximum voltage which can be applied across the capacitor leads without damaging the component). The larger the working voltage rating the more bulky and expensive the capacitor. It would appear from the graph of the power supply output (figure 3.25) that in this case, 13 volts is the worst the capacitor is likely to encounter. But consider what happens if the light bulb is disconnected from the supply. The capacitor has no current drawn from it, and will now charge right up to the peak voltage of 17 volts, and remain constantly at this voltage (figure 3.27).

Clearly we must specify a 20 volt working capacitor. The defects of this supply, namely the way that the output voltage changes if the load varies, and the fact that

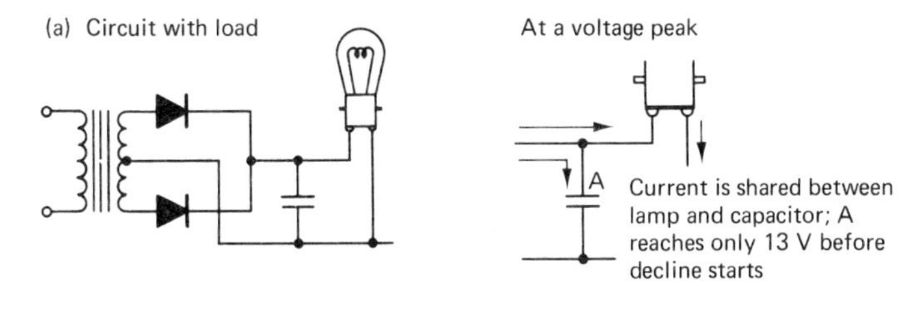


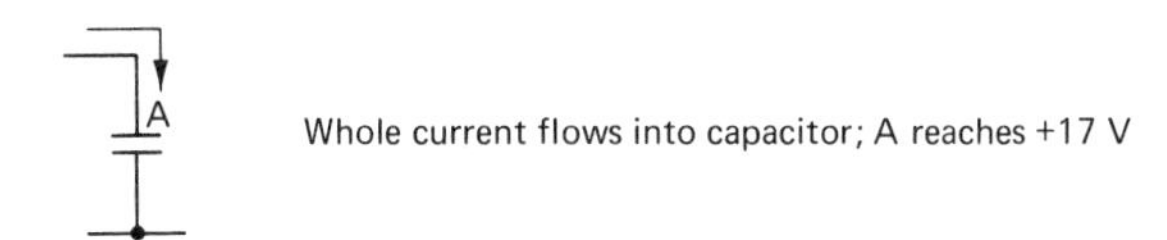


Fig. 3.27

the ripple is still considerable at its rated current, lead on to the two refinements used in most power supplies.

### The RC Low-pass Filter for Additional Smoothing

The RC high-pass filter is discussed in chapter 2 and is shown in figure 3.28(a). This AC-coupling circuit uses the capacitor's property of having a low impedance for high frequency signals and a high impedance for low frequencies to allow it to pass the high frequency components and reject the low frequency ones.

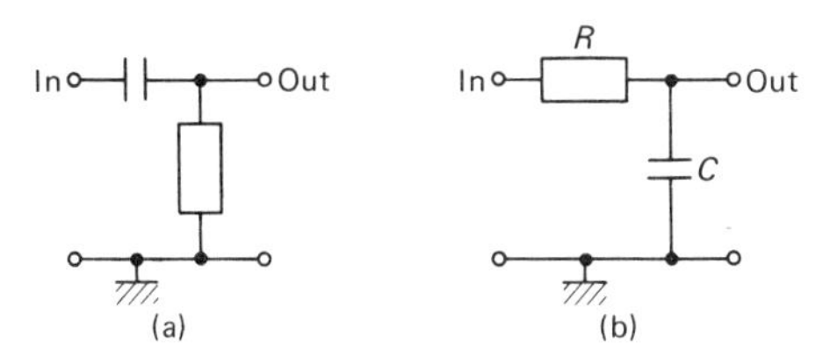


Fig. 3.28

In power-supply design we face the opposite problem – an unwanted 100 Hz 'ripple' mixed with a steady DC voltage. The solution is to use an RC low-pass filter (figure 3.28(b)). This network serves to remove the AC component by shorting it out to earth through the capacitor, which has a very high impedance to the steady DC voltage we require as an output. For this to be effective as a smoothing circuit the impedance of the capacitor at 100 Hz must be small compared to the resistance of $R$ (figure 3.29). The resistor $R$ is connected in series with the load, hence its value must be kept as small as possible to minimise the voltage across it. For instance, in the circuit shown in figure 3.30, even if $R$ is only one ohm, there will be 2 volts across it when 2 amps flow through it, and hence only 10 volts across the lamp. The only way to return the lamp to its rated 12 volts is to use a transformer with two 15 volt windings.

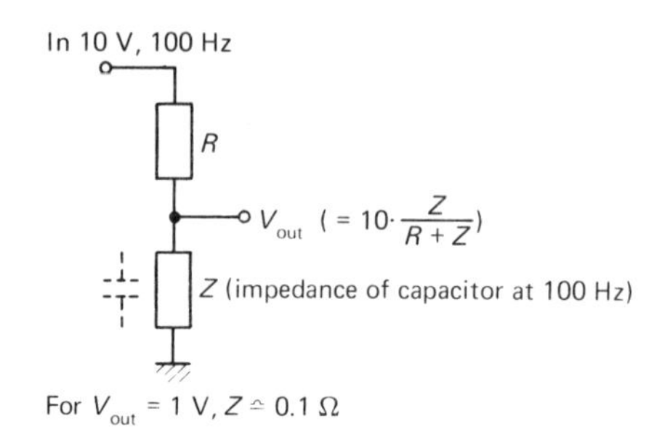


Fig. 3.29

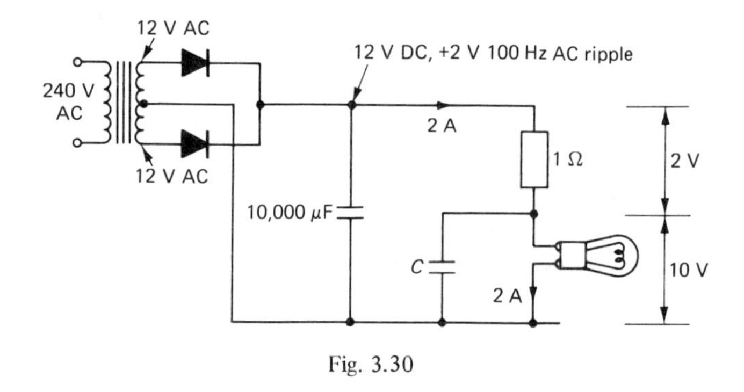


Fig. 3.30

Assuming that $R = 1\ \Omega$ and that we need to reduce the ripple from 2 V to 0.2 V, then the impedance of capacitor $C$ at 100 Hz must be 0.1 Ω. Its capacitance must then be given by the formula

$$Z = \frac{1}{2\pi FC}$$

$$\frac{1}{10} = \frac{1}{6 \times 100 \times C}$$

$$\therefore C = \frac{1}{60}\ \text{F} \approx 20\,000\ \mu\text{F}$$

The one ohm resistor needs to be capable of dissipating two watts of power.

We can now construct a suitable DC supply for a fairly powerful light source. Figure 3.31 shows the circuit details. The diodes themselves are going to get hot and need to be able to cool themselves via a metal plate called a heat sink. This can be a specially made piece of aluminium extrusion with fins, or else, more simply, one of

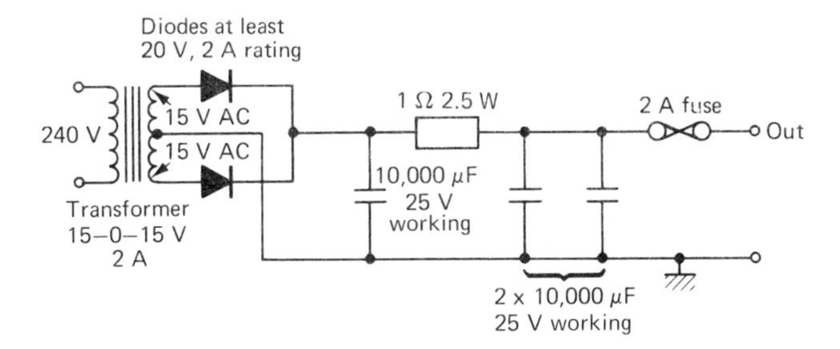


Fig. 3.31

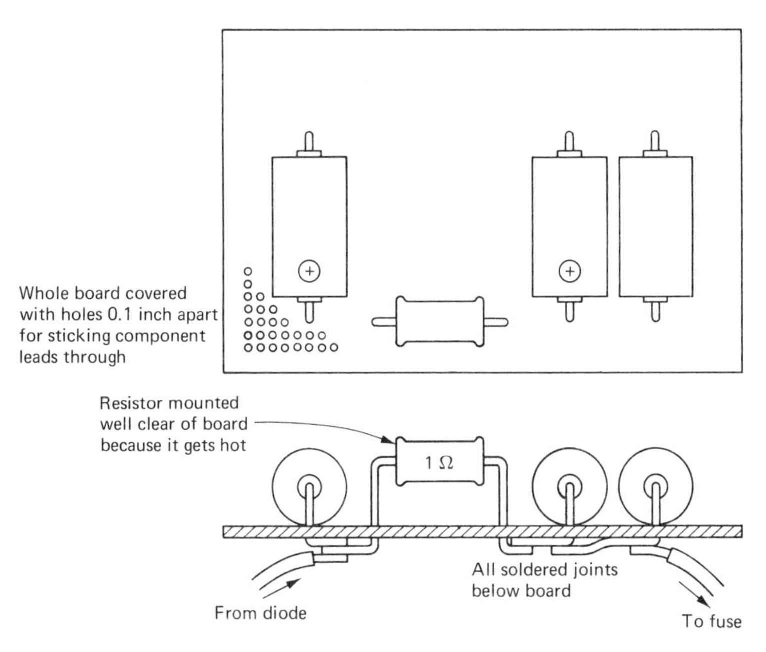


Fig. 3.32

the metal side plates of the box containing the supply. As large diodes always have one connection via their metal cases, care has to be taken to ensure that these cases do not touch the heat sink directly. It is best if the capacitors and resistors are mounted on a pre-drilled plastic board (figure 3.32).

## 3.7 REGULATION

Many electronic circuits impose a varying load on their power supplies and often will not function properly if the supply voltage alters when the load changes. To overcome this we use a device called a regulator, a component with an internal standard voltage with which it continually compares the actual output of the supply and makes appropriate adjustments by electronically controlling a resistor connected in the circuit.

Almost all the components for this complex circuit are etched on a silicon chip about 2 mm square, and encapsulated in a small metal can. A very few external components are required, mainly capacitors which are difficult to make in the integrated-circuit form. These external components are all specified in the manufacturer's literature and require no design decisions on the part of the user (figure 3.33). The regulator illustrated requires 3 volts to work – that is with a full load connected, the lowest point on the input ripple must be at least 15 volts.

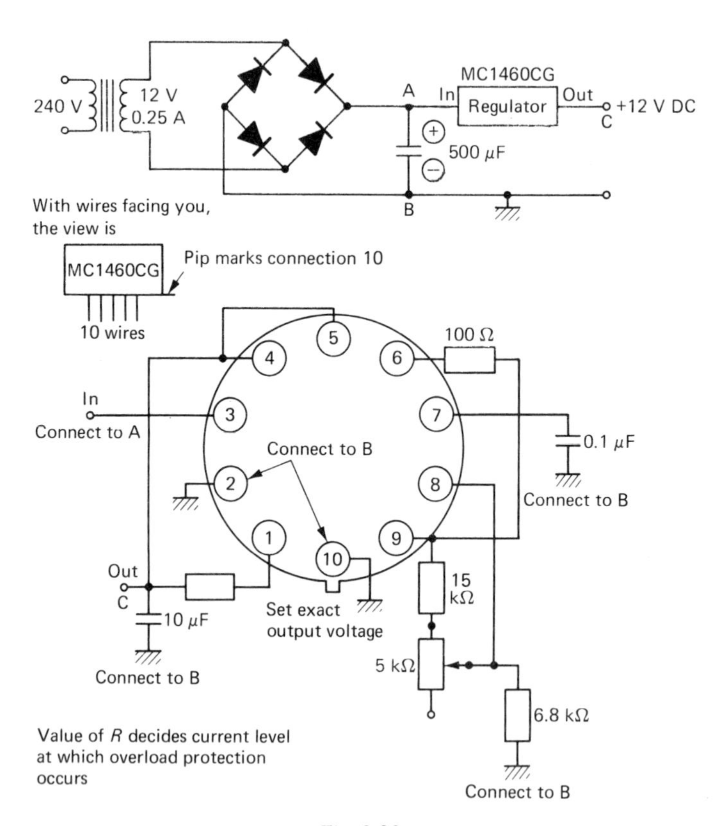


Fig. 3.33

This circuit has an added bonus of automatic overload protection so no fuse is needed. When an overload is connected the regulator internally disconnects the load and will not pass any current into the external circuit until the overload is removed.

# 4 Amplifying Small Electrical Signals

## 4.1 THE GOOD AMPLIFIER

Amplifiers increase the size of their input signal without changing its shape, as shown in figure 4.1, for example. Here, the cartridge acts as a transducer, that is a device which converts the vibrations from the wingbeats into electrical signals. But the signals are too small to work a rate meter properly so we amplify them. An amplifier used to connect a source of small signals to other circuits is often called a

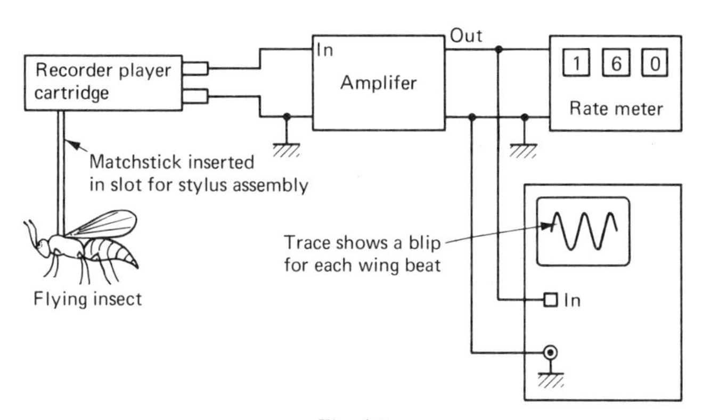


Fig. 4.1

*preamplifier*. The cartridge is the signal source for the amplifier, and as it produces signals of about 100 millivolts and the rate meter needs signals of amplitude about one volt, an amplifier with a *gain* of ten is required. The gain of an amplifier is the factor by which its output voltage is greater than its input voltage. A good amplifier has a very stable gain which is little affected by changes in its power supply voltage or the temperature of its surroundings; it does not introduce any extraneous material, it does not distort the shape of its input signal, and it takes very little current from the signal source.

**Input impedance**

All amplifiers do need to take some energy from the signal source to operate, and the amount is measured by the amplifier's *input impedance*. If an amplifier has an input impedance of 100 kΩ it means that when connected to a signal source, the amplifier will affect that signal source in the same way as a 100 kΩ resistor connected across it would (figure 4.2).

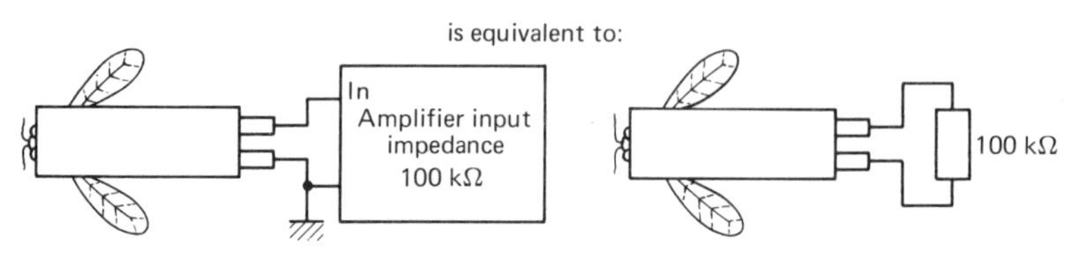


Fig. 4.2

If the pick-up cartridge is a crystal one (figure 4.3) the internal crystal element has a very high impedance (in the region of 10 MΩ). Thus what we are doing by connecting a 100 kΩ input impedance amplifier to this crystal is to set up a very disadvantageous voltage divider which loses 99 per cent of our signal before we start (figure 4.4).

If, however the record player cartridge is a magnetic one (figure 4.5) the internal element is a coil of low impedance (figure 4.6). In this case the 100 kΩ amplifier

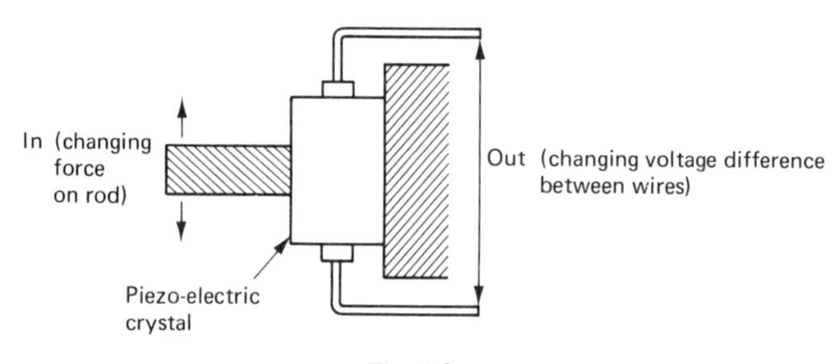


Fig. 4.3

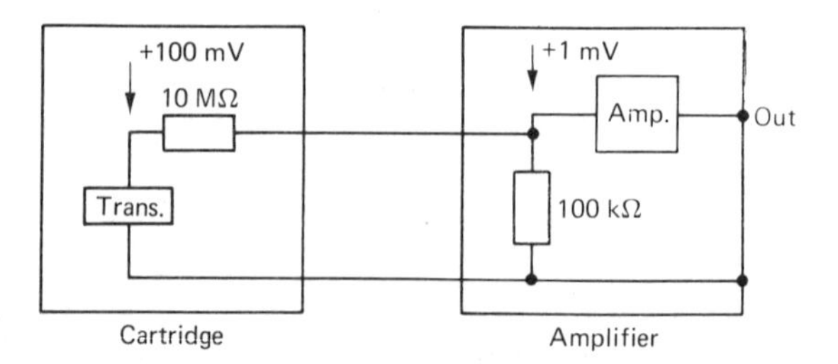


Fig. 4.4

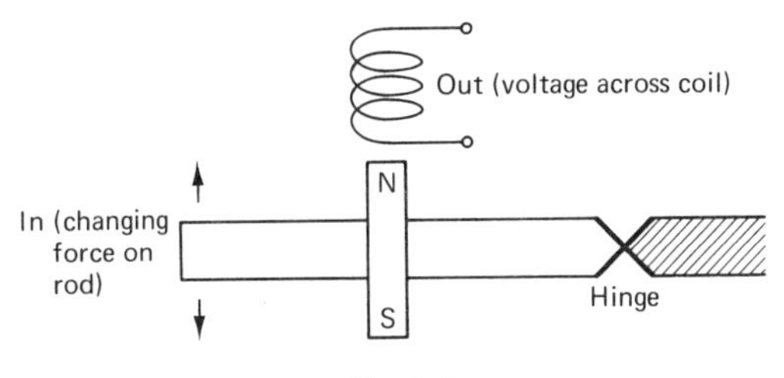


Fig. 4.5

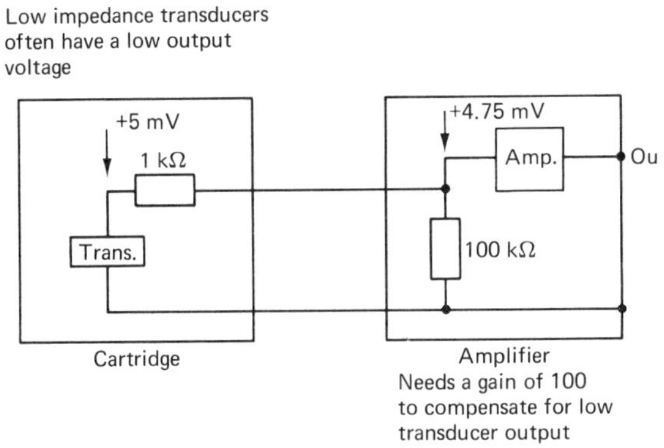

Low impedance transducers often have a low output voltage

+5 mV

1 kΩ

Trans.

Cartridge

+4.75 mV

Amp.

Ou

100 kΩ

Amplifier
Needs a gain of 100 to compensate for low transducer output

Fig. 4.6

input impedance makes only a tiny difference to the transducer output voltage. The amplifier is hence suitable for a magnetic pick-up but not for a crystal one to which it is said to be *mismatched.* As well as greatly reducing the actual output voltage of a signal source a mismatch can make it function badly and give a distorted signal, or in some cases when the signal source is an electronic circuit, can actually damage it.

## Output Impedance

An amplifier is going to act as a signal source for some other circuit or device, and to simplify impedance matching, it should have a low *output impedance*, which is

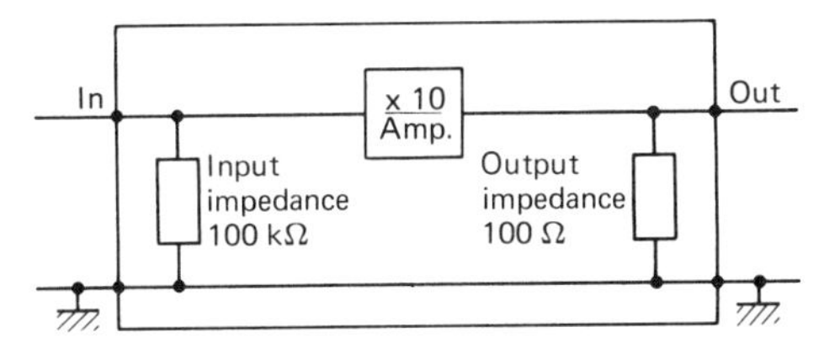


Fig. 4.7

the apparent resistance across the amplifier output terminals (figure 4.7). (It is worth making clear at this point that input and output impedances are often the impedance across the two wires of a transistor or the like, rather than actual resistors as shown in the figures.) The lower the output impedance, the more the current which can be drawn from the amplifier by whatever it is driving. In general, to ensure perfect matching the total input impedance of the circuits connected to any amplifier should be ten times its output impedance.

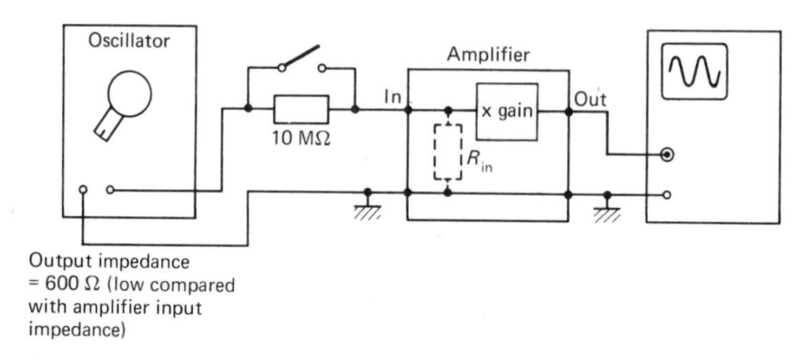


Fig. 4.8

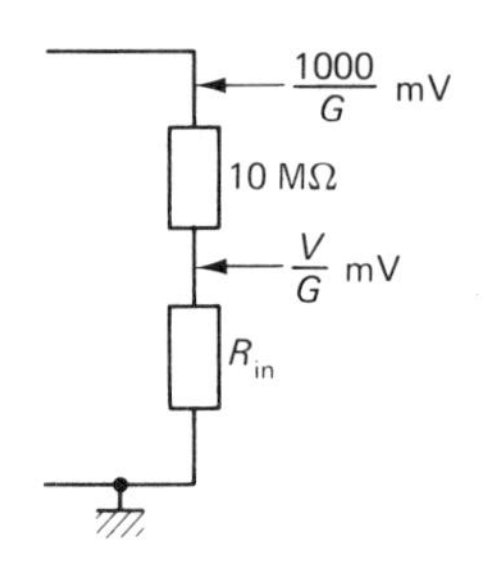


Fig. 4.9

To measure the input impedance of an amplifier (figure 4.8)

(1) Close the switch shorting out the 10 MΩ resistor and adjust the oscillator amplitude control till the oscilloscope waveform shows that the amplifier output is one volt peak-to-peak.

(2) Open switch, measure height of diminished output, $V$, in millivolts. Essentially the situation is as shown in figure 4.9.

The formula for calculating the impedance $R_{in}$ is:

$$\frac{1000}{V} = \frac{10\ \text{M}\Omega + R_{in}}{R_{in}}$$

For example, if $V = 125$ mV

Then

$$\frac{1000}{125} = 8 = \frac{10^7 + R_{in}}{R_{in}}$$

$$8R_{in} = 10^7 + R_{in}$$

$$\therefore 7R_{in} = 10^7$$

$$\therefore R_{in} = 1{\cdot}43\ \text{M}\Omega$$

## 4.2 FEEDBACK

The achievement of these requirements for good amplifier performance almost always involves using *negative feedback*. If the output of any system affects its input, the system is said to involve a feedback loop. In the system shown in figure 4.10 input is a managerial decision to declare workers redundant and the ultimate consequences

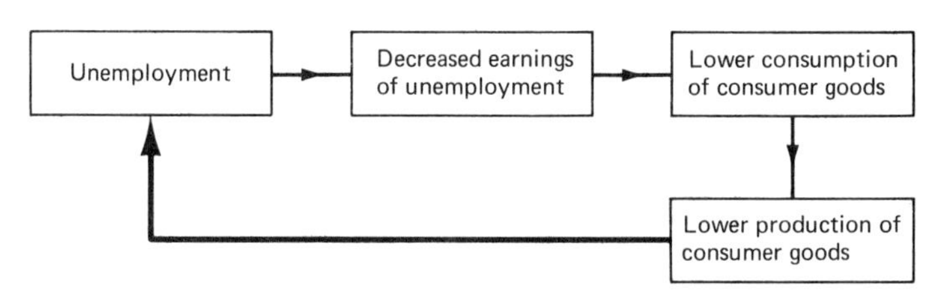


Fig. 4.10

of this decision, the lowered consumption of consumer goods, is the output of the system. This output acts as a further input to the system and acts in the same direction as the first move which set the thing going. This is called *positive feedback* or a vicious circle. Any system with positive feedback is inherently unstable because once it has started moving in a given direction the movement gathers momentum in an uncontrollable fashion, ultimately resulting in the destruction of the system concerned. For example some simple amplifiers have gains which increase as their temperature increases. If a slight overload is applied to such an amplifier it begins to dissipate too much power internally and to grow warm. This temperature rise increases the gain, so the amplifier pushes even more current into the load, dissipates more heat internally, increases its gain yet further and gets even hotter. The climax is reached in a few seconds; the amplifier blows up, fracturing its plastic encapsulation.

Positive feedback is used in many switching circuits to make the changeover between the two states of the circuit as sudden as possible. In such cases precautions are taken against things going too far.

Negative feedback occurs when the output acts on the input to oppose the direction of the initiating change (figure 4.11). Negative feedback loops are used to stabilise the output of a system, a very characteristic example being the steam engine governor (figure 4.12). If the engine goes faster, for example if the load it is

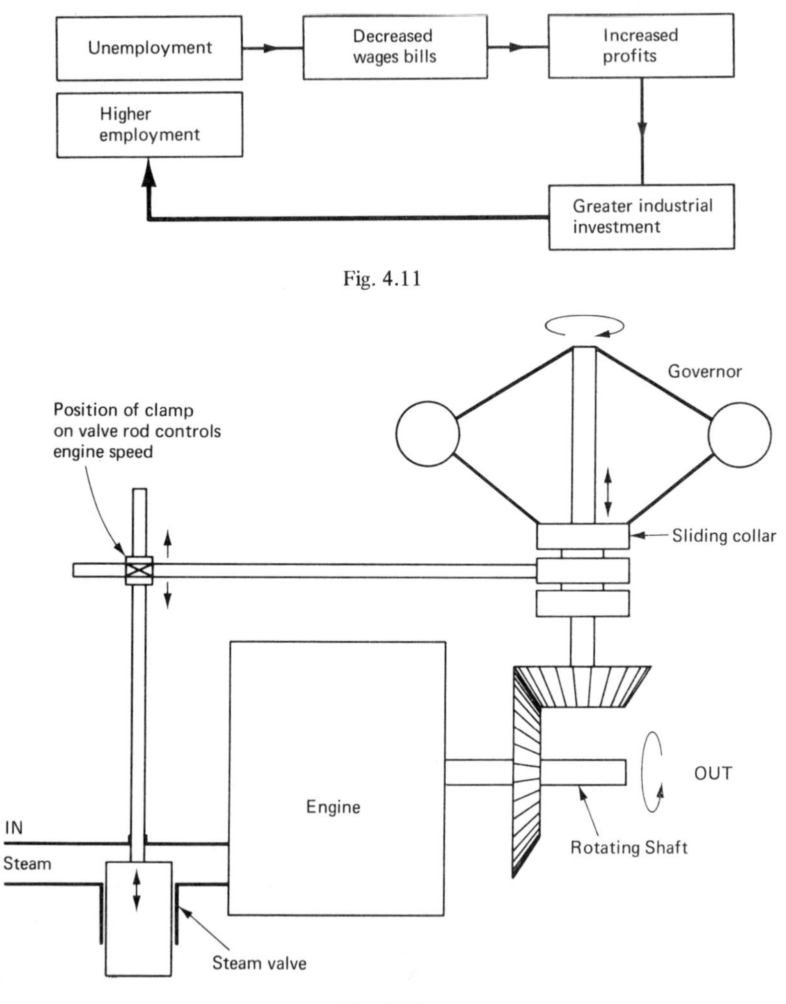


Fig. 4.11

Fig. 4.12

driving becomes lighter, the balls of the governor fly out, raising the collar, hence reducing the steam supply and slowing the engine down. If the shaft slows, the governor balls move inwards, lowering the sliding collar and increasing both the steam supply and hence the engine speed. This engine will thus run at a constant speed whatever its load, and the speed required can be set by changing the position at which the rod from the governor is clamped to the steam valve.

## 4.3 CIRCUITS USING ELECTRONIC NEGATIVE FEEDBACK

### The Operational Amplifier

Operational amplifiers are so-called because they were first developed to perform mathematical operations in analogue computers in the 1940's. Now they are available very cheaply as single integrated-circuit (IC) components, and are rapidly becoming ubiquitous. The symbol (figure 4.13(a))

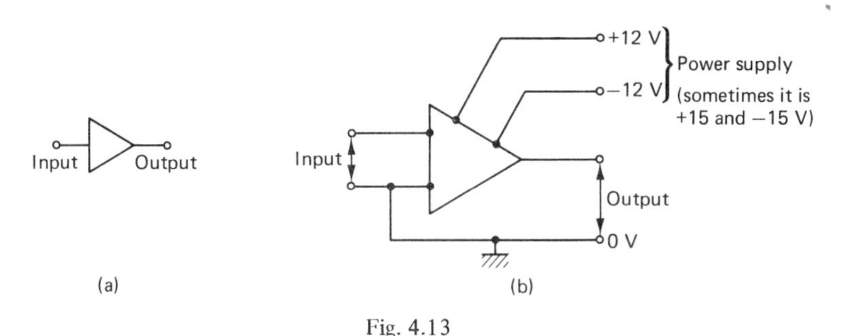


Fig. 4.13

is a shorthand form, omitting the power supply and earth connections since they always remain the same. Figure 4.13b gives a more complete representation. A small change in the voltage at the input causes a very large change going in the opposite direction at the output (figure 4.14). A typical integrated-circuit

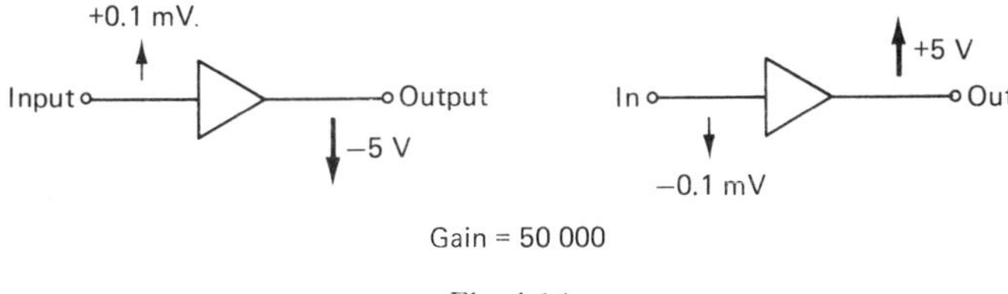


Fig. 4.14

operational amplifier has a gain (output voltage)/(input voltage) of 50 000, an input impedance of 400 kΩ and an output impedance of 150 Ω.

It is very unusual to attempt to use an operational amplifier to increase the amplitude of a signal by 50 000 times. More commonly the gain is severely limited by a feedback resistor connecting the amplifier output and input (figure 4.15). Consider a current just beginning to flow along wire into the amplifier itself. The output voltage of the amplifier will start to move sharply in a negative direction dragging the current up wire Y and through R instead. When this trend starts going too far and current just begins to flow out of the amplifier input towards resistor *R*, the amplifier output voltage at once begins to move in a positive-going direction, preventing this 'overshoot' from getting anywhere.

The feedback loop thus prevents the point labelled 'In' ever moving more than a tiny fraction of a volt away from zero volts (if the gain of the amplifier is 50 000, and its output is capable of moving between + 5 and − 5 volts, its input is constrained between − 5/50 000 = − 0.1 mV and + 5/50 000 = + 0.1 mV). Any input current is channelled straight round the feedback resistor (figure 4.16). In

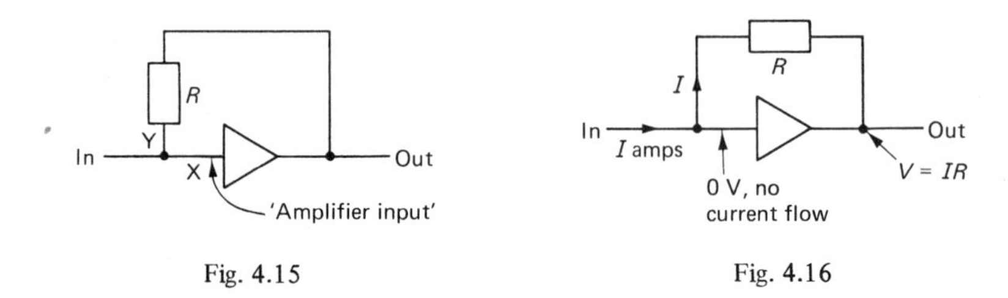


Fig. 4.15

Fig. 4.16

practice, when designing operational amplifier circuits one does not think a great deal about feedback loops, but simply remembers the two rules:

(1) the amplifier input terminal always stays at zero volts. (It is sometimes, I think misleadingly, called a *virtual earth*.)

(2) all the input current goes round the feedback loop, and none into the amplifer itself.

### The Current Amplifier

An obvious application of the operational amplifier is to measure a current. A typical instance is the use of a cadmium sulphide photocell, a device whose resistance depends on the light intensity falling on it (figure 4.17). In the cell illustrated, the resistance between A and B decreases as the light intensity falling on the cell increases. All the photocell current, $I$ amps, flows through the feedback resistor producing an output voltage $V = I \times 10^6$. (The feedback resistor is 1 MΩ, i.e. $10^6$ ohms.) So a meter reading of 1 volt indicates that the photocell current = $1/10^6$ amps = 1 microamp, and hence the photocell resistance is given by

$$R = \frac{12 \text{ (volts)}}{10^{-6} \text{ (amps)}} = 12 \text{ M}\Omega$$

### Construction

The cheapest IC operational amplifiers require three external components to set their frequency response. In this case the complete circuit is given in figure 4.18. IC operational amplifiers come either in small cylindrical metal cans ('TO100') or black plastic rectangles ('dual in line'). Many biologists prefer to TO100 cans, because they are more rugged, you can cram them closer together on the plastic board used for building circuits, and they are (arguably) easier to solder to. Figure 4.19 shows the details of the pin connections for an operational amplifier. The last

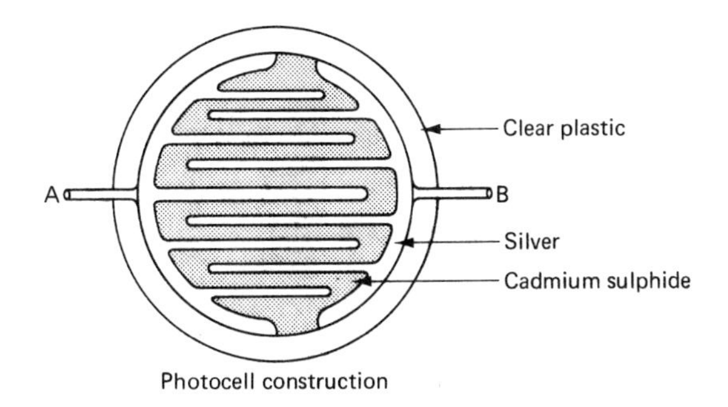


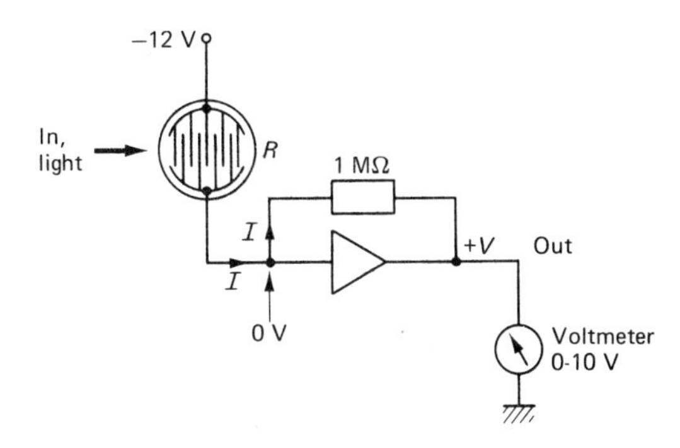


Fig. 4.17

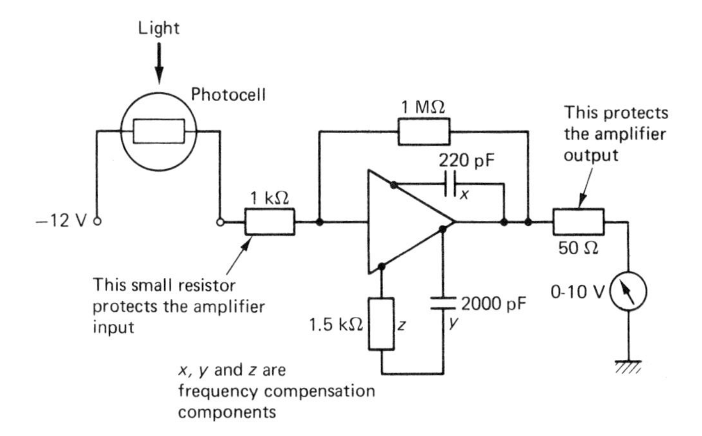


Fig. 4.18

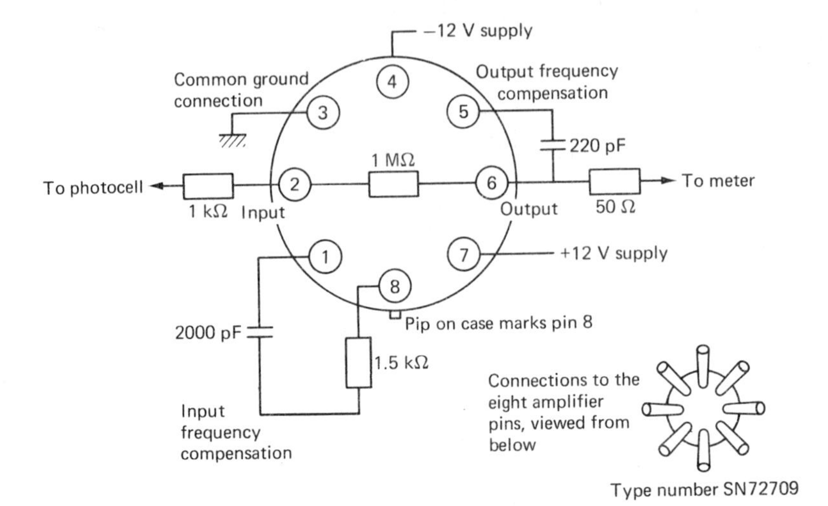


Fig. 4.19

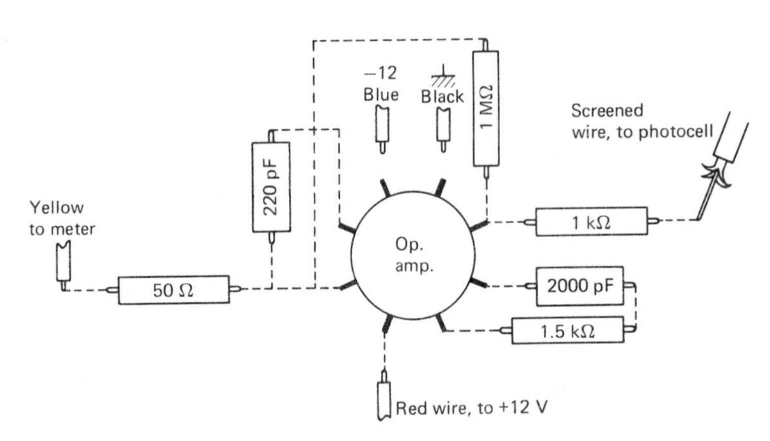


Fig. 4.20

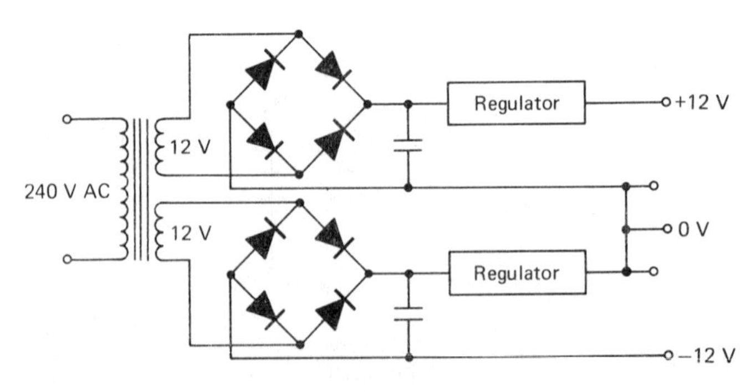


Fig. 4.21

three digits of the number (709 in this case) indicate the amplifier type and would be the same for identical amplifiers made by different firms. Figure 4.20 illustrates the constructional details, viewing the circuit board from above. The connections underneath the board are shown dotted.

Regarding power supplies, avoid any method of making do with a single 24 volt power supply; a useful supply design is given in figure 4.21.

The circuit in figure 4.19 also provides for frequency compensation. This is designed to cut down the gain of the amplifier at high frequencies and thus prevent it oscillating (producing a steady output of some high frequency without any input being connected). This is provided by the components labelled $R_1$, $C_1$, $C_2$ in figure 4.22. The values to be used depend on the ratio of the feedback resistor to the

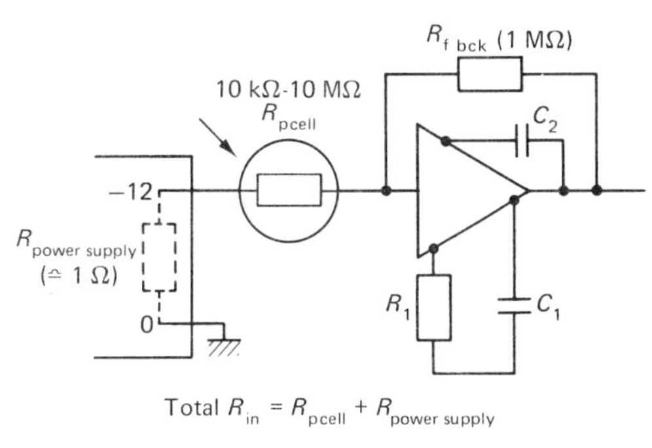


Fig. 4.22

resistance between the input terminal and ground as table 4.1 shows. In the present case photocell $R$ is variable, so it is largely a matter of guesswork. As high frequency response is not important in this circuit I have used fairly large capacitors which definitely prevent the amplifier oscillating. The operational amplifier shown in figure 4.22 is a – 709 series one; the more expensive – 741 series amplifiers do not need external frequency compensation.

Table 4.1

| $R_{f\ bck}$ / Total $R_{in}$ | $R_1$ (k Ω) | $C_1$ (pF) | $C_2$ (pF) |
|---|---|---|---|
| 1 | 1.5 | 5000 | 220 |
| 10 | 1.5 | 500 | 20 |
| 100 | 1.5 | 100 | 3 |

A constant problem with amplifiers is the presence of unwanted 50 Hz signals from the AC mains. This is often called 'hum' because of its characteristic sound when played through a loudspeaker. It is caused by mains leads, every one of which can act as a radio transmitter broadcasting 50 Hz signals; every exposed wire is capable of picking them up. To avoid this happening it is necessary to enclose in an earthed sheath all wires longer than 20 mm or so which are connected to the amplifier input. This is done by using coaxial cable. Figure 4.23 shows the construction of this cable, and also the correct screening for the photocell amplifier, assuming that the photocell is to go on a long lead which plugs into a box containing

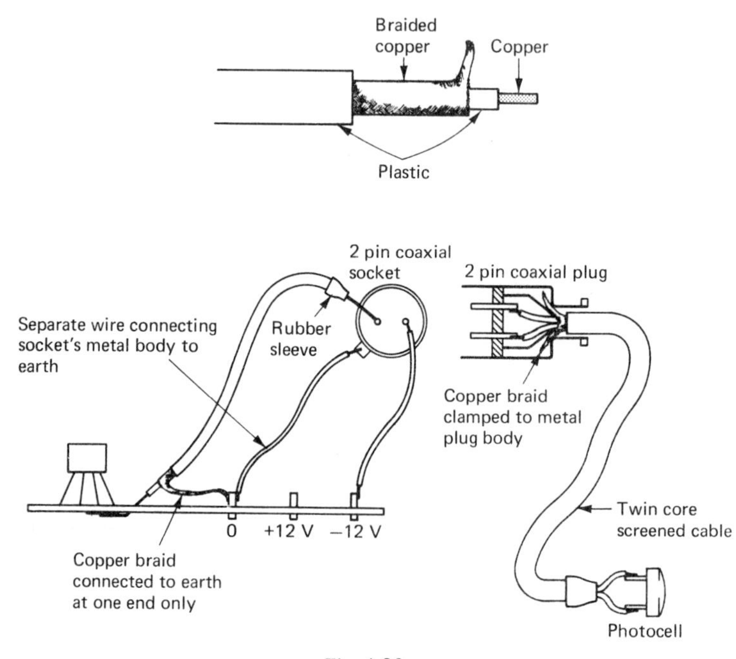


Fig. 4.23

the amplifier. Photocell circuits are often difficult to rid of hum. If problems arise it is worth remembering that fluorescent strip lights flash 100 times per second, and that even filament lights running on mains do cool appreciably between each half cycle, thus having a light output with 100 Hz ripple. Any photocell measuring light from such a source will tend to reproduce this variation in light intensity, though the problem is less severe with CdS cells which respond rather slowly to changes in light intensity.

## The Voltage Amplifier

(*i*) An input resistor converts a voltage to be amplified to a current suitable for driving the current amplifier described in the last section (figure 4.24). If the input of this system is at $x$ volts, we can apply Ohm's Law to the input resistor:

$$I = \frac{x}{R_{\text{in}}}$$

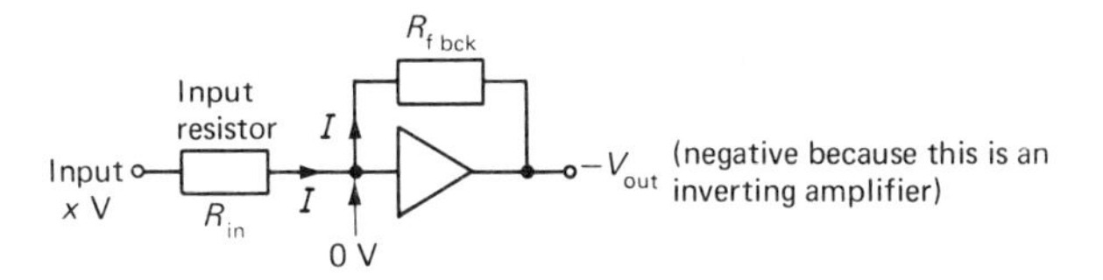


Fig. 4.24

All the input current must then go round the feedback resistor, and applying Ohm's Law to this

$$I = \frac{V_{out}}{R_{fbck}}$$

so

$$\frac{x}{R_{in}} = \frac{V_{out}}{R_{fbck}}$$

and

$$V_{out} = x \cdot \frac{R_{fbck}}{R_{in}}$$

The gain of this amplifier depends on the ratio of two resistors and so is very easy to specify exactly, and also very independent of ambient temperature variation. Figure 4.25 shows an amplifier with an input impedance = $R_{in}$ = 100 kΩ and an output impedance = 150 Ω. The feedback resistor has a value of 1 MΩ, and thus the gain is ten. It is exactly ten for signals of all frequencies from DC to 100 kHz, then gradually falls to zero at about 10 MHz.

(*ii*) *Amplifier with variable gain.* Adjusting either the input resistor or the feedback resistor alters the gain of the amplifier. Increasing $R_{in}$ decreases the gain, and increasing $R_{fbck}$ increases it. But changing the input resistor changes the input impedance too, so it is best left alone.

One solution is simply a variable feedback resistor (figure 4.26). An alternative arrangement is shown in figure 4.27; here $R_x$ and $R_y$ act as a voltage divider. The voltage at P is the amplifier output voltage reduced by the ratio of the resistors:

$$\text{Voltage at P} = (\text{Output voltage}) \times \frac{R_y}{R_x + R_y}$$

The effect of attenuating the output voltage before connecting it to the feedback resistor is the same as using a much bigger feedback resistor:

$$R_{fbck} = R_f \times \frac{R_x + R_y}{R_y}$$

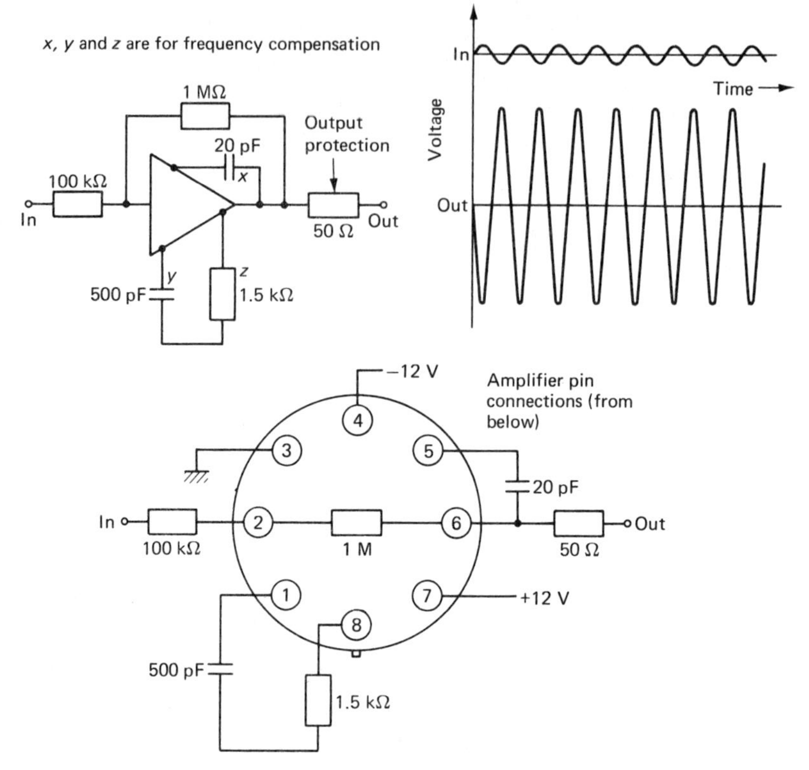


Fig. 4.25

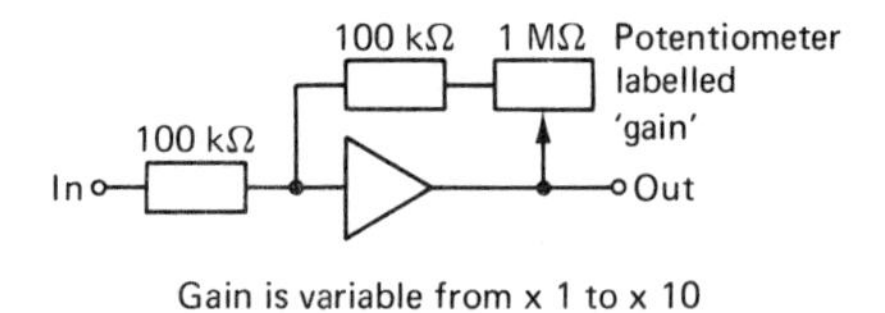


Fig. 4.26

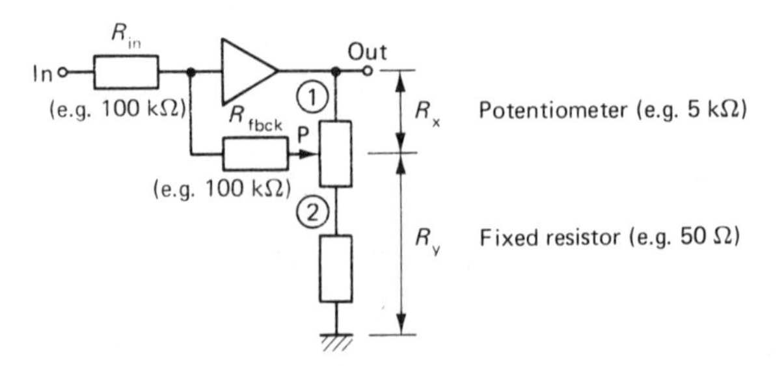


Fig. 4.27

Here, $R_{fbck}$ is the equivalent feedback resistor. If $R_y$ is very small the feedback resistor becomes enormously large and the amplifier approaches its total gain of 50 000 and becomes very unstable. Hence the use of a fixed resistor to limit the extent to which $R_y$ can be reduced, and thus set a reasonable maximum gain. For example, taking the component values shown in brackets in figure 4.27, when the potentiometer is set in position 1, the feedback resistor is 100 kΩ and the gain is 1. In position 2 the effective feedback resistor is

$$100\ \text{k}\Omega \times \frac{5050}{50} \simeq 10\ \text{M}\Omega$$

so the gain is 100.

If feedback resistors larger than 10 MΩ are used with –709 type operational amplifiers the resulting amplifiers tend to suffer from drift, a very slow change in the output voltage without any change in the input.

(*iii*) *The adder.* It is possible to have a voltage amplifier with more than one input resistor. An example of the type of circuit used is shown in figure 4.28, and it is useful for adding two signals together:

$$V_{out} = 10\,(V_{ina} + V_{inb})$$

Because the feedback loop keeps the junction of the two input resistors at zero volts there is no tendency for $I_a$ to flow back up $R_{inb}$ and vice versa, so there is no possibility of the sources of the two signals interacting, a problem frequently met when mixing two signals. There is no need for the input resistors to have the same value (figure 4.29).

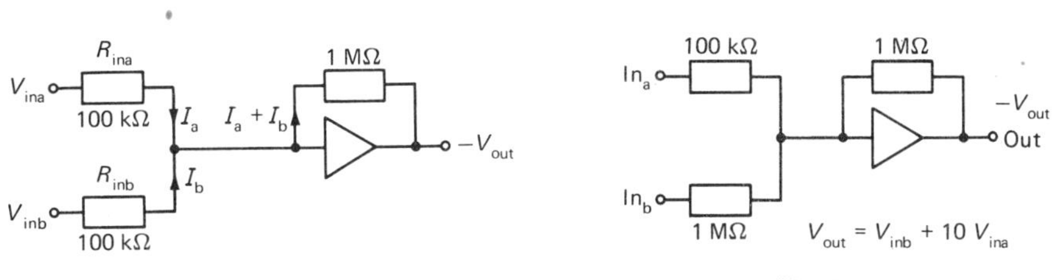


Fig. 4.28 Fig. 4.29

A common use of the adder is for *backing off*, that is subtracting a constant voltage from a signal. The need for this occurs, for instance, when you are trying to use a photocell to measure a slight dimming in a bright light. The bright light will cause a large steady current to flow through the photocell, and, if the amplifier is sensitive enough to detect small changes (i.e. has a large enough feedback resistor), this large photocell current is going to result in an output voltage larger than the amplifier can cope with (figure 4.30). In fact the amplifier cannot manage output voltages greater than + 11 or – 11 volts, and simply stays at – 11 volts for all input currents which should result in output voltages more negative than this. Hence no change in the output will result from a slight dimming of the light. The solution is shown in figure 4.31. The potentiometer supplies an adjustable steady voltage between zero and − 12, and is adjusted so that the contribution to the amplifier input from the 100 kΩ input resistor is − 24 μA, just enough to cancel out the photocell current due to the steady light.

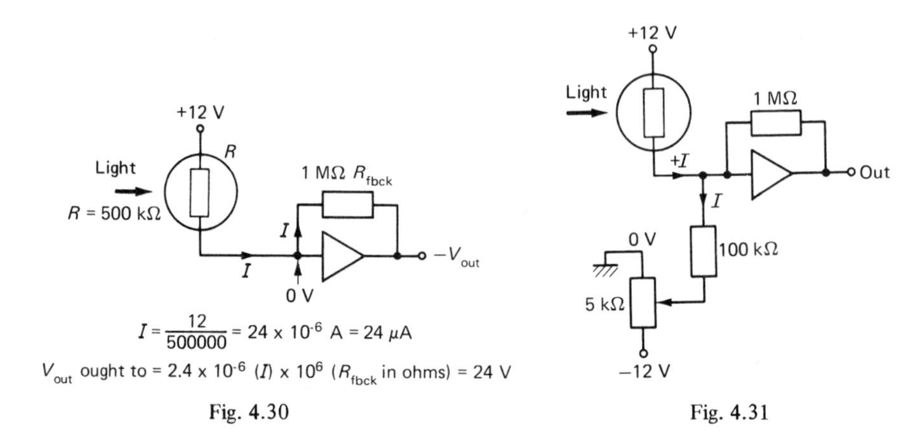


Fig. 4.30

Fig. 4.31

The amplifier output is now zero volts for the steady light, and will become slightly negative for a small dimming. As an additional bonus the steady light level can be read from a scale calibrated on the knob adjusting the potentiometer.

## The Voltage Follower

(*i*) *Unity gain follower.* The ground connection on the operational amplifier can also be used as input. It is sometimes called the 'non-inverting' input, because when it is used the output voltage moves in the same direction as the input voltage. The voltage follower circuit which uses this non-inverting input has a very high input impedance (around 100 MΩ) a gain of one ('unity gain'), and a low output impedance (150 Ω). It is non-inverting (figure 4.32). At first sight an amplifier whose output is completely indistinguishable from its input seems singularly useless. But in fact it is the perfect answer to all impedance matching problems, because it can be used to give any electronic device a very high input impedance without having any other effect on its behaviour. For instance, in the case of the 10 MΩ crystal pick-up cartridge mentioned at the beginning of this chapter figure 4.33 would be a very suitable circuit.

The voltage follower circuit's action depends on the way in which the two inputs are interconnected, see–saw fashion. In figure 4.34, pushing on the see–saw corresponds to an input of positive voltage, and pulling an input of negative voltage. All the circuits we have considered so far have had the non-inverting input connected to zero volts (neither pushed nor pulled) and a large feedback resistor (corresponding to a very long pointer) has prevented the inverting input from moving more than a fraction of a volt from zero. In other words the see–saw has barely shuddered.

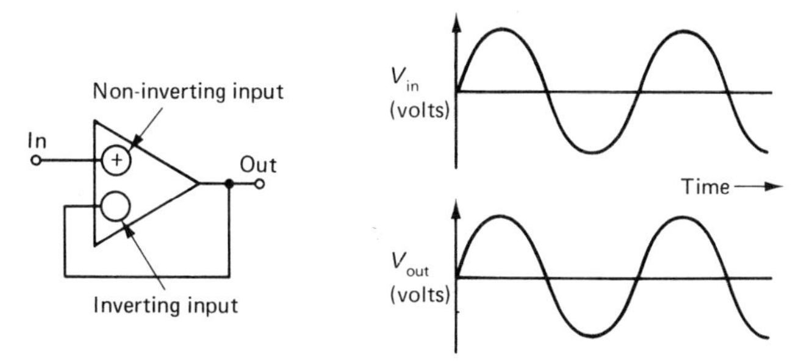


The ground connections are to the power supply:

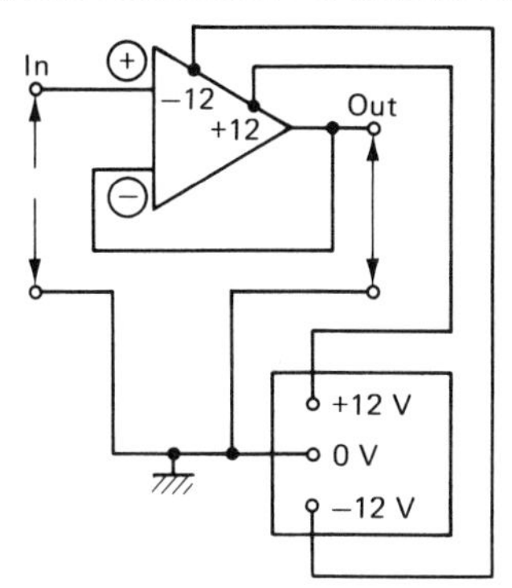


Fig. 4.32

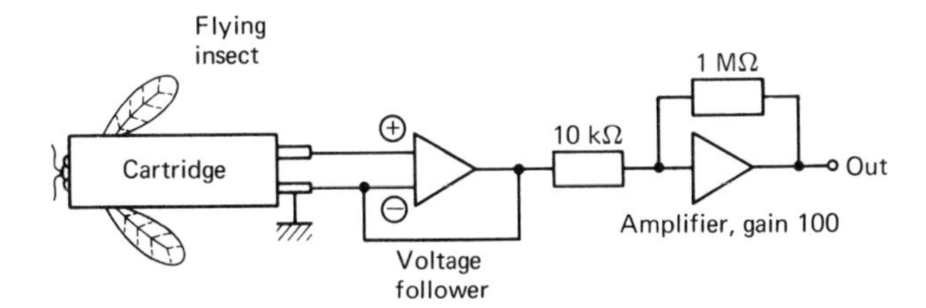


Fig. 4.33

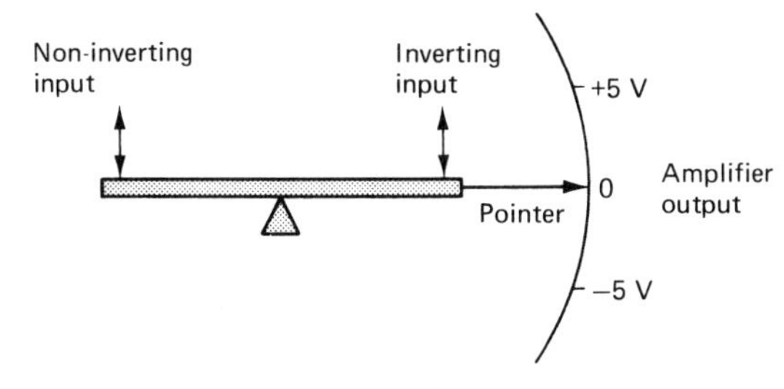


Fig. 4.34

In this circuit the feedback resistor is zero, so the pointer length is very short, and the see–saw moves vigorously as the amplifier output changes. Pushing down on one end of a see–saw causes the other end to push upward. Similarly making a current flow into the inverting input of an operational amplifier connected as a voltage follower means that the same current will emerge from the non-inverting input. Consider a positive input voltage of + 2 volts beginning to be applied to the non-inverting input of the voltage follower. The output will begin to move rapidly in a positive direction, so current will begin flowing into the inverting input, promptly to re-emerge from non-inverting input, trying to push back the initial input current. As soon as the output and inverting input exceed + 2 volts the current trying to get into the non-inverting input will be beaten back, there will be a net flow of current out of this input, and the amplifier output voltage will fall. So the negative feedback loop maintains the voltage at the output and the inverting input at exactly the voltage applied to the non-inverting input, and prevents any current flowing into or out of the non-inverting input. Hence the very high input impedance which results.

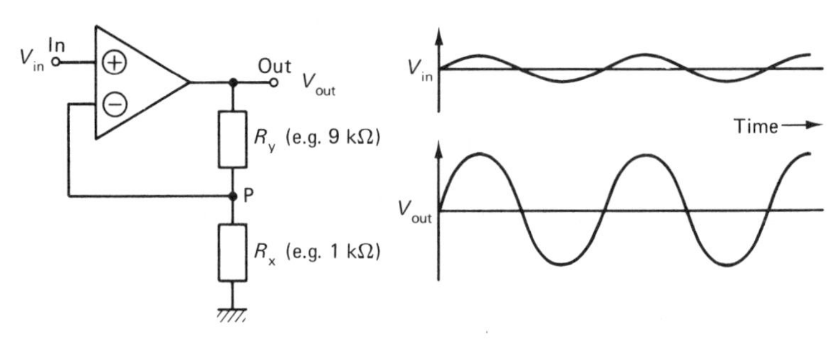


Fig. 4.35

(*ii*) *Voltage follower with gain* (*figure 4.35*). The negative feedback loop, as before, maintains the voltage at the inverting input of precisely the level of that at the non-inverting input. Hence the voltage at point P = input voltage, $V_{in}$. Now

$$\text{Voltage at P} = V_{out} \times \left(\frac{R_X}{R_X + R_Y}\right)$$

so

$$V_{out} = V_{in} \times \left(\frac{R_X + R_Y}{R_X}\right)$$

The gain of the amplifier = $(R_X + R_Y)/R_X$, so the component values shown in brackets constitute a ×10 voltage follower.

(*iii*) *Differential amplifier.* This circuit is a hybrid between the adjustable gain inverting amplifier and the voltage follower with gain (figure 4.36). The formula for the gain is obtained by applying the basic principles:

(1) no current flows into the amplifier input;
(2) the voltage at the + ve input = voltage at – ve input;
(3) the currents round a T-junction add up:

$i_1 \quad i_2 \quad i_3$

$$i_1 = i_2 + i_3$$

and by applying Ohm's law to each resistor. To make the amplifier gain variable, make $R_X$ a variable resistor.

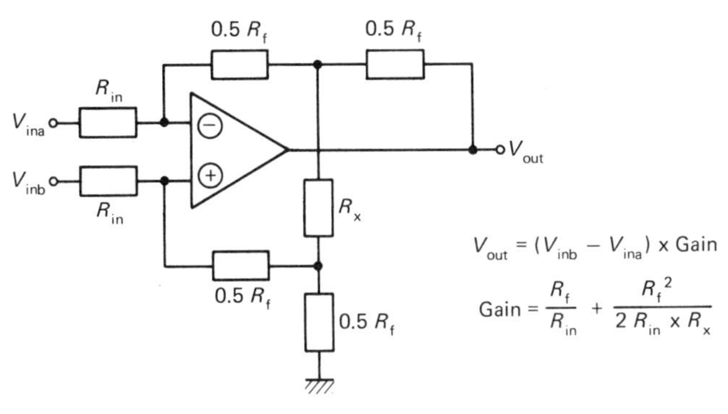


Fig. 4.36

This amplifier responds only to differences between the two inputs. If, say, a + 1 volt signal is connected to both inputs no output at all will result. This property, called rejection of the common-mode signal, is useful for removing interference picked up by leads if you are trying, for instance, to record the voltage across a cell membrane (figure 4.37). Provided both leads follow similar routes to the

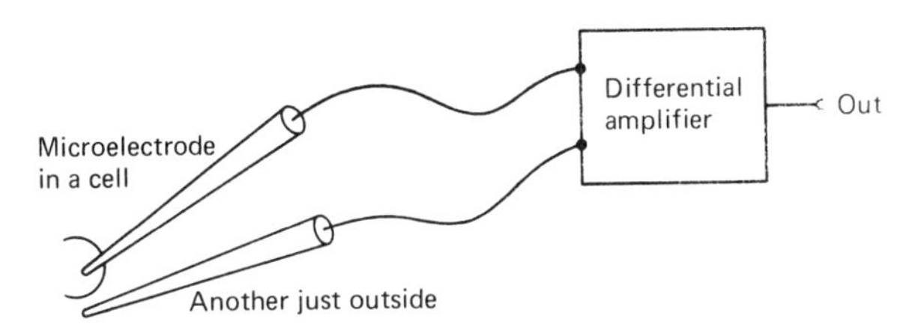


Fig. 4.37

amplifier they will pick up identical hum, which will be ignored, while the voltage across the membrane will represent a difference between the inputs and will be amplified.

## 4.4 NOISE AND THE PHASE-SENSITIVE AMPLIFIER

Two of the principal reasons for unsatisfactory amplifier operation, hum (unwanted 50 Hz signal) and drift (unwanted very slow signals), have been discussed earlier. The third major problem is noise, or unwanted random signals. This makes a constantly changing spikey pattern on the oscilloscope, and is responsible for the hissing noise heard when a VHF radio is not tuned in.

It is a very intractable problem because it is largely due to the thermal motions of electrons. Its level is set by the simple formula

$$\text{Noise (volts)} = \sqrt{(kTBR)}$$

where $k$ is Boltzmann's constant, $T$ is the absolute temperature in $K$, $B$ is the bandwidth of the amplifier in Hertz and $R$ is the source impedance in ohms. Boltzmann's

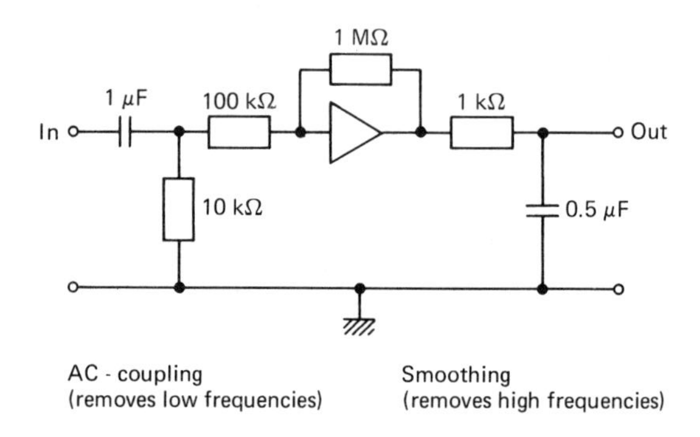


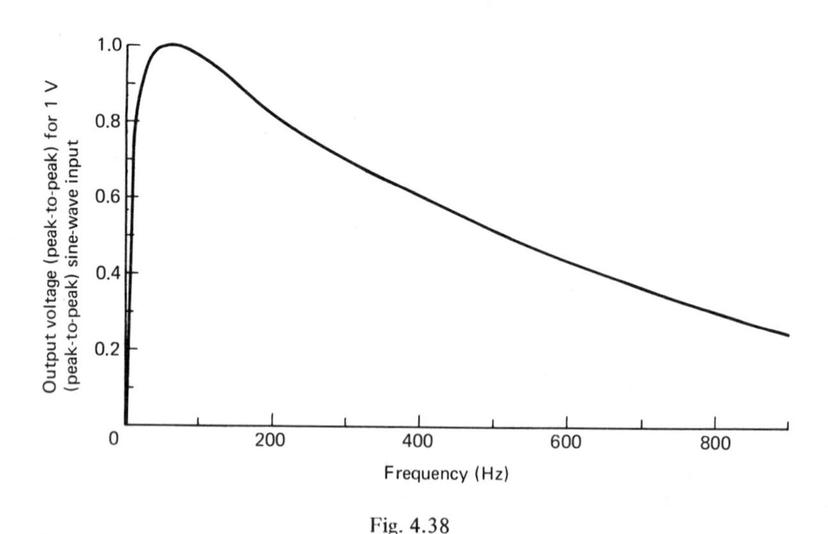


Fig. 4.38

constant and the impedance of the source are immutable. The temperature of the signal source is also often difficult to alter, though sometime photocells are put into liquid nitrogen. To take a typical case if one attempts to record from a nerve with a microelectrode of impedance 40 MΩ (about normal) and an amplifier bandwidth (gap between lowest and highest frequencies to which amplifier responds) from DC to 10 kHz, then

$$V_{noise} = \sqrt{(10^{-23} \times 1.38) \times (10^4) \times (4 \times 10^7) \times (3 \times 10^2)}$$
$$= \sqrt{1.65 \times 10^{-9}}$$
$$= 10^{-5} \times \sqrt{16.5}$$
$$\approx 40 \times 10^{-6} \text{ volts}$$
$$\approx 40\ \mu\text{V}$$

Hence it is most unlikely that it will be possible to detect reliably any voltages in the nerve of less than 80 μvolts. The only way to overcome noise is to reduce the amplifier bandwidth, disconnecting unwanted slow signals by AC-coupling and unwanted fast signals by smoothing the amplifier output (figure 4.38). There is a limit to how far you can curtail an amplifier's bandwidth without beginning to discard relevant signals too, and unfortunately this limit is frequently reached before the noise has been reduced to a satisfactory level.

One approach which is sometimes useful in these intractable cases is to use the phase-sensitive amplifier, which only accepts input signals labelled by being pulsed at a known frequency. For example, the input signal in a flying insect detector is the slight dimming in an infra-red light when an insect flies through the gap between the source itself and a photocell. If the light source is a solid-state lamp it can be switched on and off very rapidly and its light output can easily be pulsed at a rate of 2500 Hz by using a square-wave oscillator as its power source. The signal from the photocell is amplified with a current amplifier (phase-sensitive amplifiers work badly on very small input signals), then fed into the phase-sensitive amplifier (figure 4.39). This consists first of an AC-coupling circuit which rejects all the DC and a high proportion of the low frequency signals from the photocell (figure 4.40). We are in a position to take this step because we know that all the relevant signals are going to be at a frequency of 2.5 kHz, the lamp flash rate, so all that this circuit rejects is bound to be noise.

Next comes the electronic switch, a component called a field-effect transistor or FET, and another amplifier (figure 4.41). The resistance between the drain and source of a FET is controlled by the gate voltage (figures 4.42 and 4.43). When the light is off, output 1 of the oscillator is at + 12 volts, so the FET is in its 'switch closed' state (figure 4.44). The amplifier's input is shorted to ground, and no signal from the photocell can get through. When the light is on, output 1 of the oscillator is at − 12 volts, and the FET is in the 'switch open' condition (figure 4.45). Signals from the photocell can now get through to the amplifier.

The field-effect transistor is particularly suitable for this kind of switch because it has a very high impedance between drain and source when 'off' and because it does not matter whether the signals it is switching are positive or negative voltages (figure 4.46). Either the drain or the source will behave as the source, depending on

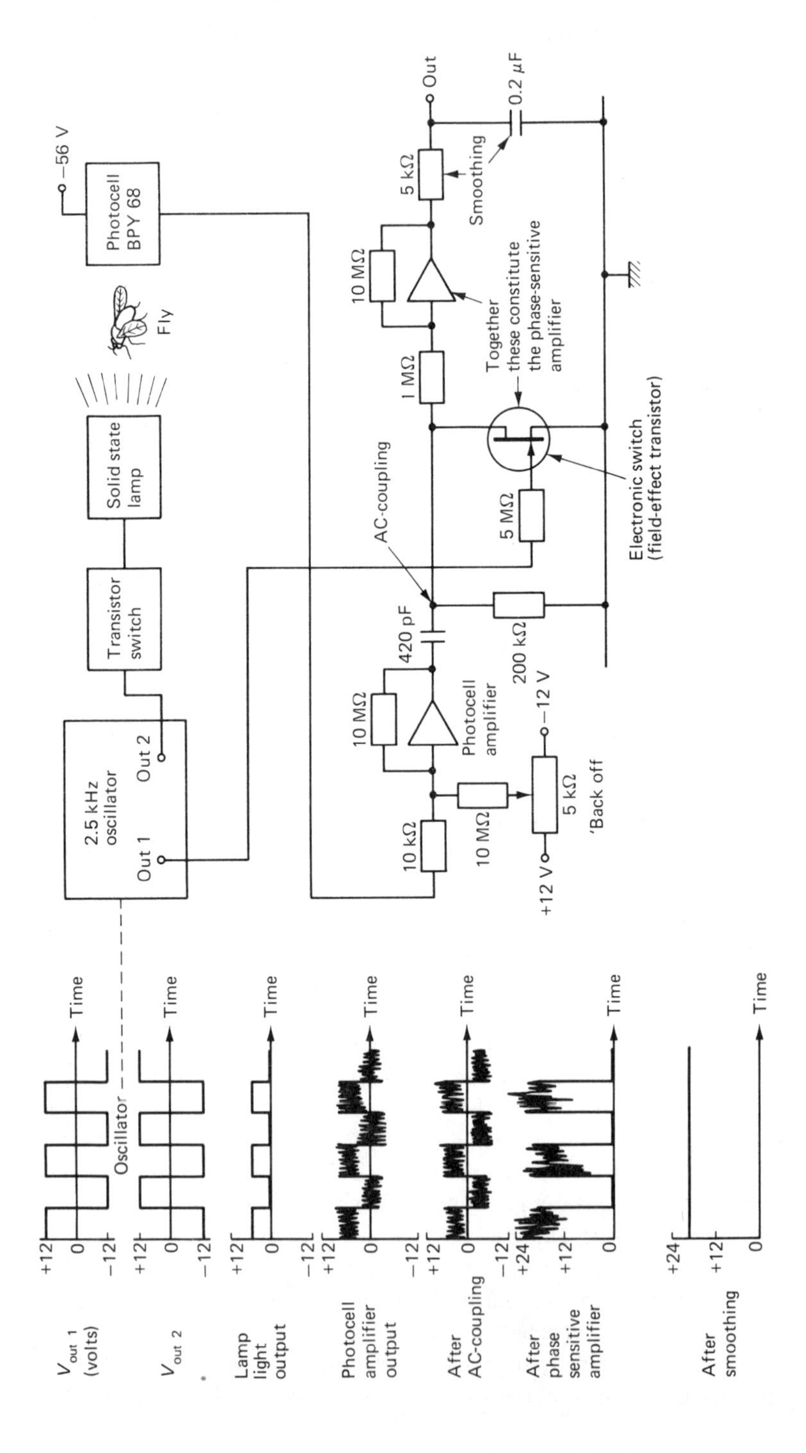

−56 V
Photocell BPY 68
Fly
Solid state lamp
Transistor switch
2.5 kHz oscillator
Out 1
Out 2
Out
0.2 μF
Smoothing
5 kΩ
10 MΩ
Together these constitute the phase-sensitive amplifier
1 MΩ
Electronic switch (field-effect transistor)
5 MΩ
AC-coupling
420 pF
200 kΩ
10 MΩ
Photocell amplifier
−12 V
10 kΩ
10 MΩ
5 kΩ
'Back off
+12 V
$V_{out\ 1}$ (volts)
$V_{out\ 2}$
Lamp light output
Photocell amplifier output
After AC-coupling
After phase sensitive amplifier
After smoothing
Oscillator
Time
+12
0
−12
+24

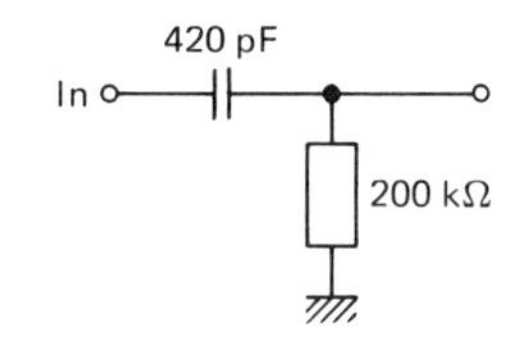


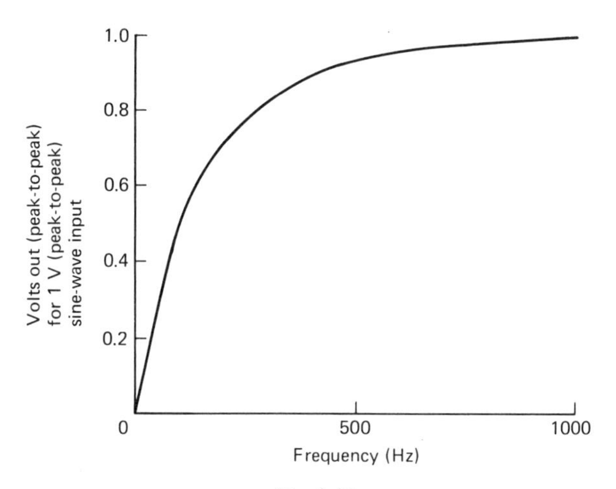


Fig. 4.40

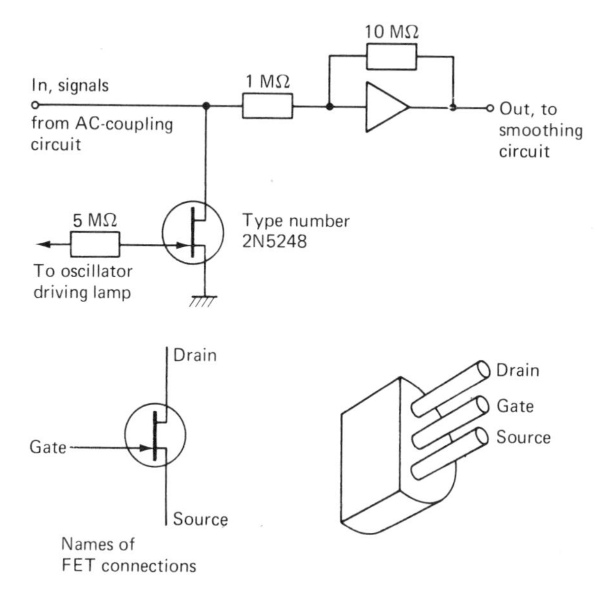


Fig. 4.41

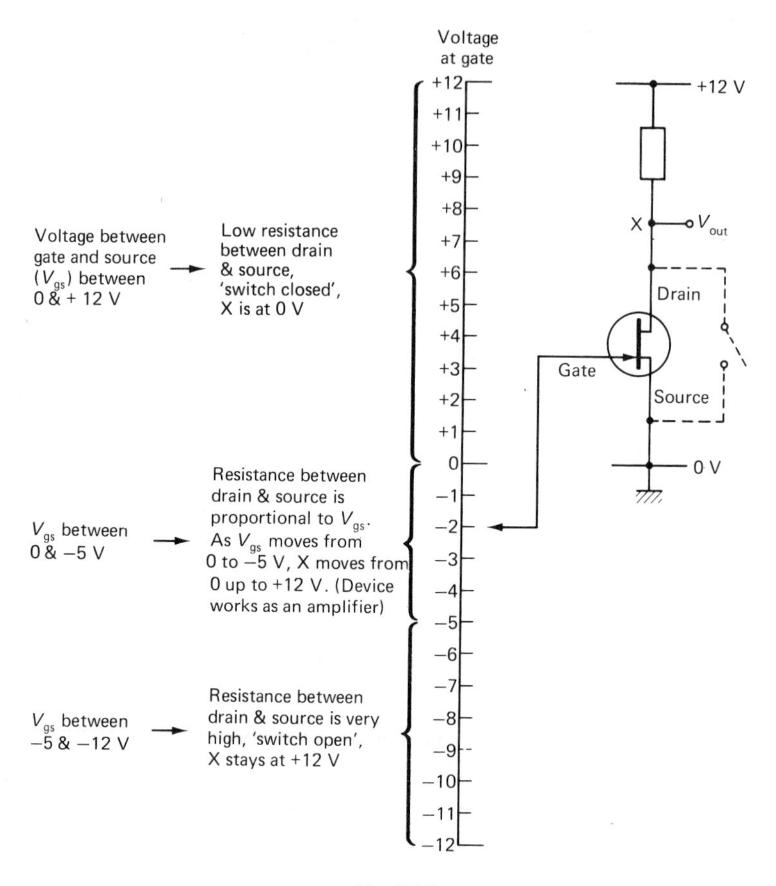


Fig. 4.42

which of the two is the more negative. Even in the most difficult case, in which the switch is supposed to be 'off', and the maximum negative signal is applied, $V_{gs}$ at 5 volts is still enough to ensure that the switch stays off (see figure 4.42). FET's have an enormous resistance provided the gate is at a voltage negative to the source voltage. It is as if they had a diode connecting the gate and the source (figure 4.47). The 5 MΩ input resistor is used in this circuit to protect the transistor from big input currents when the input voltage from the oscillator is at + 12 volts. It has no effect at all when the input voltage is − 12.

The last stage of the phase-sensitive amplifier circuit is a smoothing capacitor (figure 4.48). This circuit smooths out fairly vigorously any signals above 250 Hz -- one tenth of the lamp flash rate. Only signals originating from the lamp can affect this output voltage. Signals of a higher frequency will spend as much time negative as positive during the times when the switch is open (figure 4.49). Smoothing circuits give the average voltage, proportional to total area above the line (shaded)

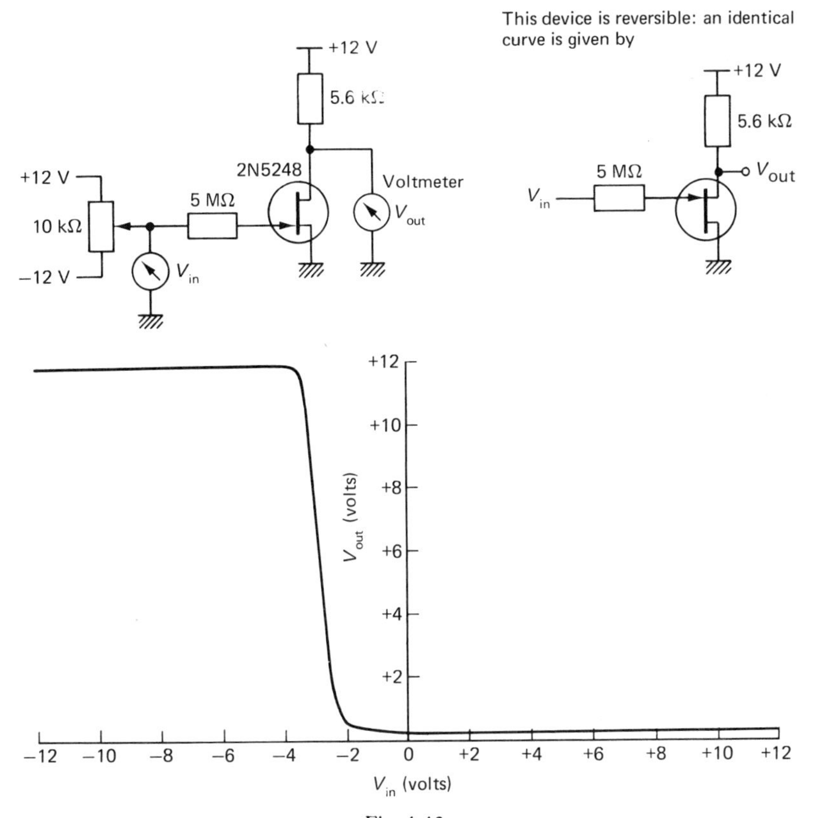


Fig. 4.43

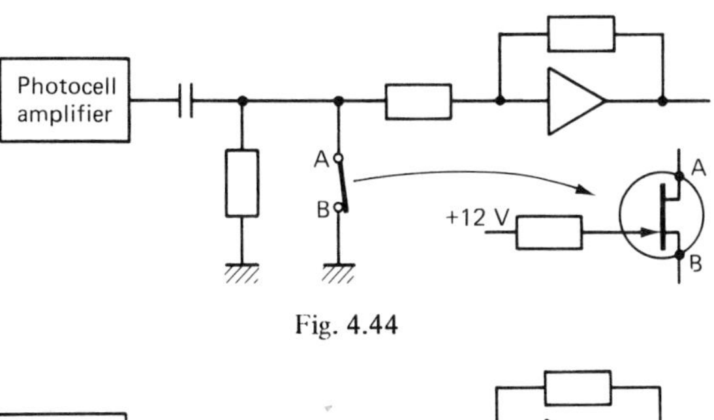


Fig. 4.44

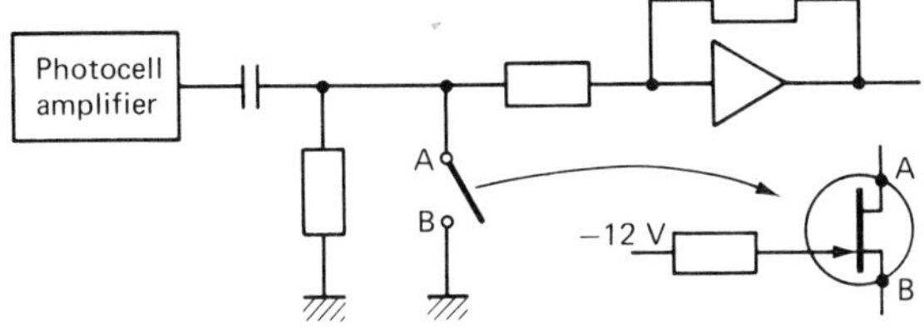


Fig. 4.45

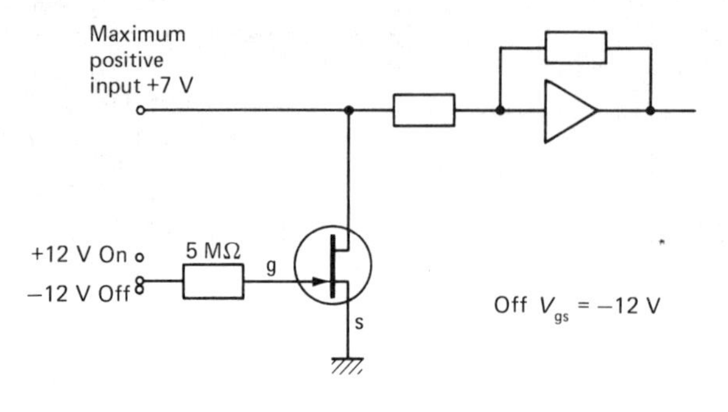


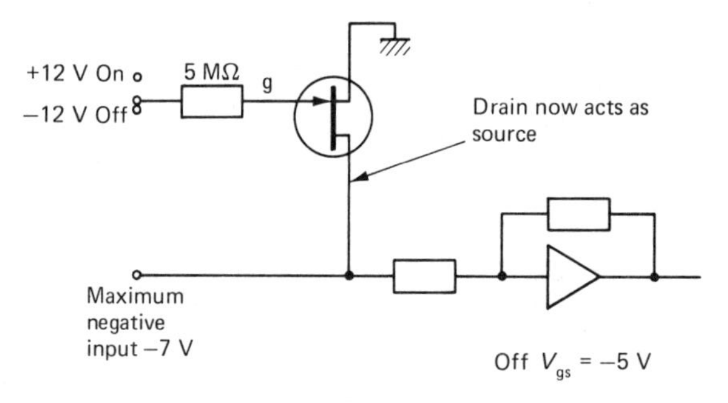


Fig. 4.46

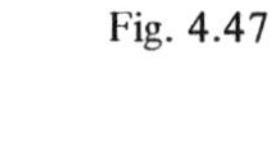


Fig. 4.47

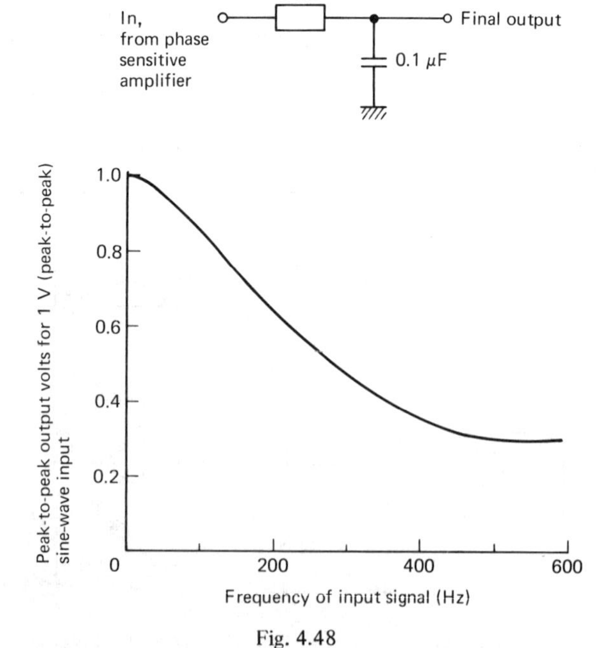


Fig. 4.48

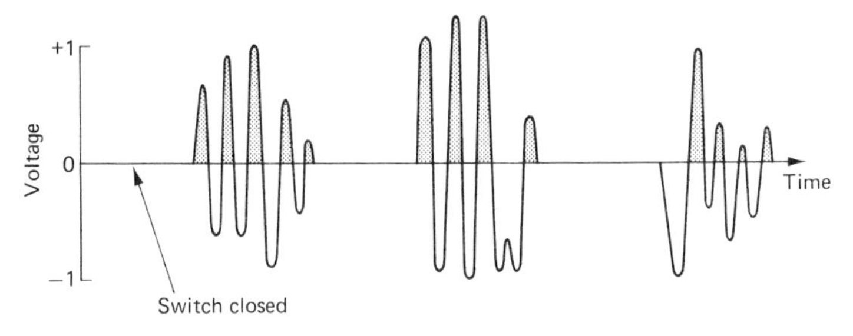


Fig. 4.49

minus the unshaded area below the line, which equals zero. Even signals at the same frequency as the lamp, but with each cycle not starting at the right time ('out of phase') tend not to produce an output voltage (figure 4.50). The smoother averages out over about ten of these sampling periods, so the only thing that could conceivably affect the output voltage (other than signals from the lamp) is noise in the range DC to 250 Hz. But we have already removed all of this with the AC-coupler before the switch, and hence our final output is a steady voltage proportional to the light

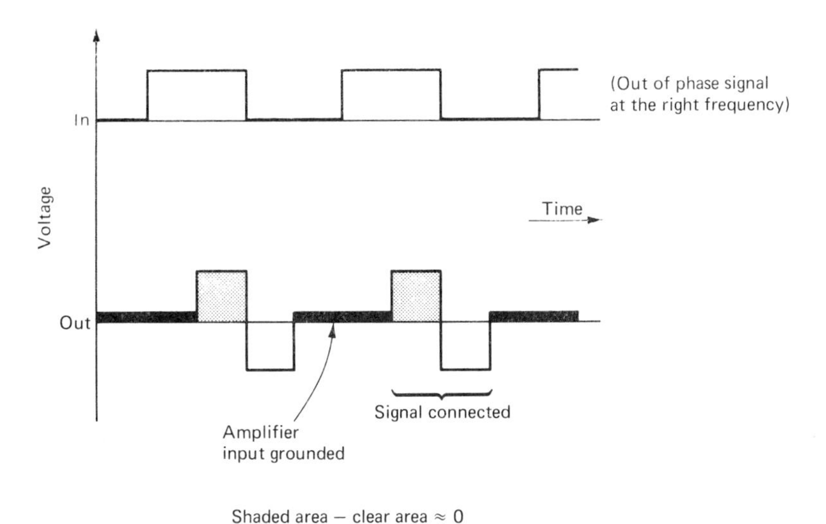


Shaded area − clear area ≈ 0

Fig. 4.50

intensity from the flashing stimulus lamp falling on the photocell. This circuit is unresponsive to changes in ambient illumination (but if these are large it may be advisable to AC-couple the preamplifier input to avoid the amplifier being driven hard over).

This circuit involves changing a steady light intensity into a series of flashes (figure 4.51(a)); transforming the electrical signal from the photocell representing these flashes into an alternating waveform (b); then cutting off the bits below the

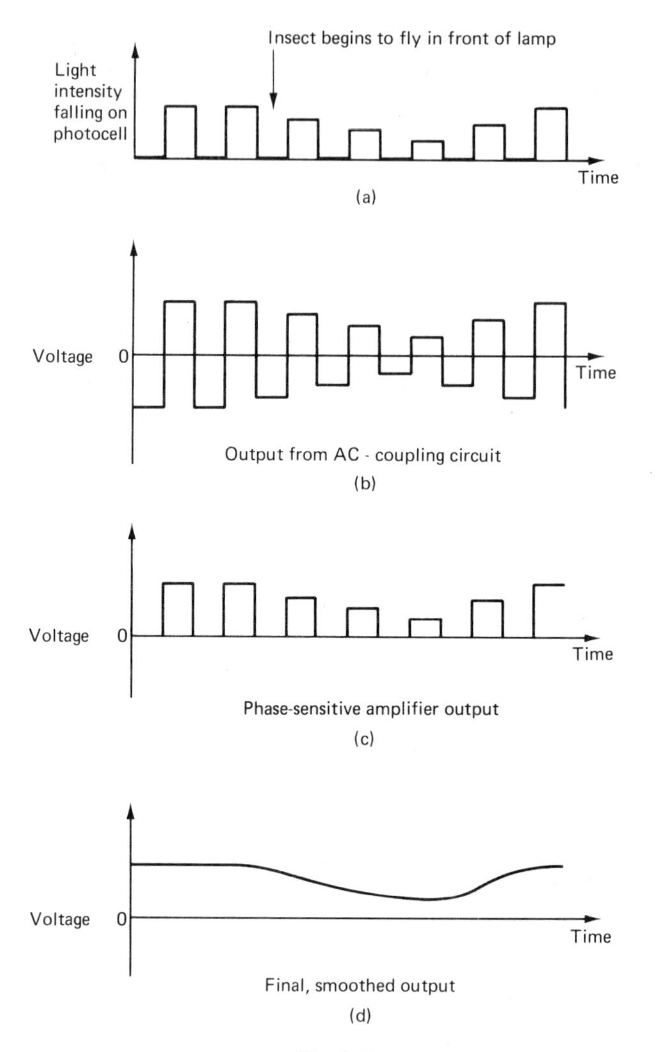


Fig. 4.51

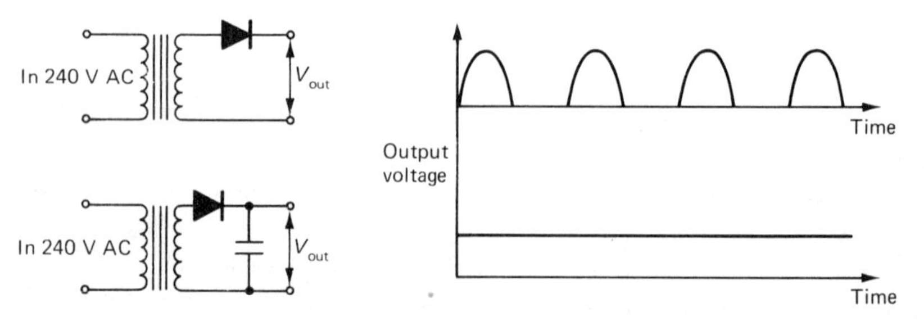


Fig. 4.52

zero volt line with a switch worked from the flashing lamp supply (c) to get back a steady voltage proportional to the intensity of the lamp if it had not been flashed to start with (d).

The last two stages are very similar to the events in a mains power supply (figure 4.52) except that in this case we have an intelligent diode which only lets voltages pass when it knows that the signal might conceivably be there (i.e. when the lamp is on) rather than the power supply diode which simply lets everything positive through.

This circuit, sometimes called a phase-sensitive rectifier because of this analogy, obviously loses three-quarters of the potential signal voltage from the photocell (which is halved because the lamp is off half the time, and halved again by the switch disconnecting the negative portions of the signal after AC-coupling) but, as it also loses up to 99.9 per cent of the noise, is of enormous help in improving the signal/noise ratio (the measure of how well an amplifier works in noisy conditions).

## 4.5 PHYSIOLOGICAL AMPLIFIERS

Electrical impulses can be detected in nerve cells if the cell is impaled with an electrode, called the recording electrode, and the voltage difference between this and a second electrode – the indifferent electrode – is very greatly amplified. The electrodes are most commonly glass tubes, drawn to very fine tips in the region of one micron diameter, and filled with a saline solution. (See chapter 6 for more details.) These electrodes have a very high impedance, of around 50 MΩ. Hence the amplifier used must have a very high input impedance, both to avoid reducing the already tiny signal voltage available at the electrode tip by an unfavourable voltage divider, as shown in figure 4.4, and also to minimise any interference with the normal functioning of the nerve cell. The highest input impedances, of up to $10^{17}$ ohms, are obtained by using a high gain, high input impedance operational amplifier in a voltage follower circuit. The Ancom type 15a–17a amplifier is particularly suitable because it has two sets of inputs linked to the same amplifier. One pair is used for the electrodes, and the other for the feedback loop. Best results are obtained with a unity gain voltage follower driving a standard voltage amplifier which provides the necessary gain (figure 4.53).

Amplifiers sold specifically as biological amplifiers use similar principles, but compromise a little on the very high input impedance in order to provide extremely low noise levels and rapid recovery from traumatic overloads. This last is particularly important when the nerve being recorded from is stimulated by an electrical impulse. The stimulating impulse is applied between two stimulating electrodes, one implanted in the nerve cord and one in neighbouring tissue. These are kept as far as possible from the recording electrodes to minimise the extent to which the stimulus pulse is applied directly across the amplifier input, causing a large output signal known as the stimulus artefact. A small stimulus artefact can be very useful as a marker on the record, enabling the latency of the nerve spike it triggers to be easily measured, but large stimulus artefacts drive the amplifier hard over, and it often does not recover before the nerve spike arrives. If this problem persists despite careful screening of all the leads to electrodes, the trouble is probably that the indifferent recording electrode and one of the stimulating electrodes are in some way electrically connected, most likely via earth connections. Problems sometimes arise even when the stimulator apparently has no earth connections at all. In this case the tortuous path to earth involves a capacitor made by the primary and secondary windings of the stimulator's mains transformer. The solution to these stubborn cases of large stimulus artefacts is to use *radio-frequency coupling* between the stimulator box and the stimulating electrodes. The stimulator pulse supplies

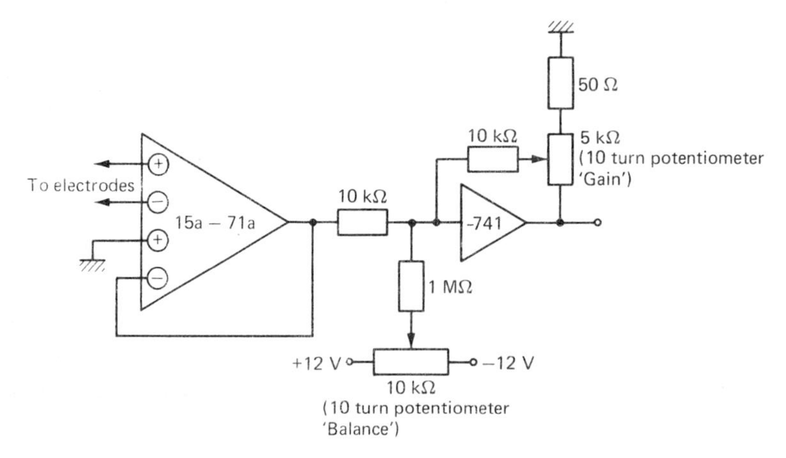


Fig. 4.53

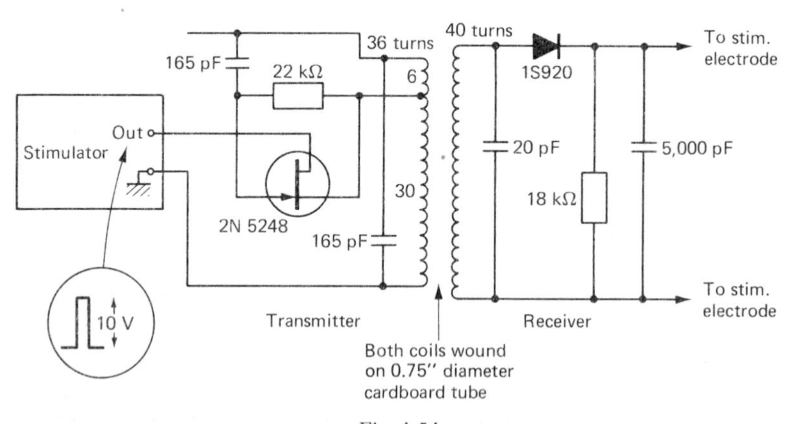


Fig. 4.54

the power to operate a miniature radio transmitter, which has a small coil of wire as a transmitting aerial (antenna). On top of this is wound a receiving aerial coil, which is connected to the stimulating electrodes via a traditional crystal set circuit to reconstitute the original waveform (figure 4.54).

In general high-gain high input impedance amplifiers connected to biological preparations are extremely prone to hum and interference due to sparks in ill-suppressed switches, and great care must be taken with screening. It is best to arrange that at least the animal and the preamplifier (and, preferably, the experimenter as well) are in a steel mesh cage connected to a good earth. The cladding on the cable which carried the main electricity supply into the whole building is often adequate as an earth; otherwise bury a large copper pipe in moist earth and bring a fat wire specially to your set-up. All earth connections should go to one point, screened cables should have connections to their screens at one end only, and no part of the apparatus should have more than one wire leading to earth. 'Low microphonic' (no spurious signals when you move the wire) coaxial cable, which contains a layer of black electrically conducting plastic underneath the standard plaited copper sheath, greatly improves leads from the electrodes to the preamplifier. To minimise hum pick-up it is important that no mains leads should pass inside the screened cage. A common culprit is the lead to the microscope lamp, which must be run from a DC supply.

# 5 Datalogging

## 5.1 DATA-REDUCTION

Recording data in an experiment can be either inclusive, for example taking a colour cine film while observing two sticklebacks fighting, or reductive, for example observing the same fight stopwatch in hand and noting the times and duration of the various bouts of displacement activity. The first method enables you to relive the experiment exactly. The second involves some of the work of testing your theory in the act of acquiring data. A typical example involving electrical signals would be an experiment using an electrode to record the activity of a group of nerve cells (figure 5.1). The experimenter wants to know about the activity of cell B during certain one-minute experimental periods.

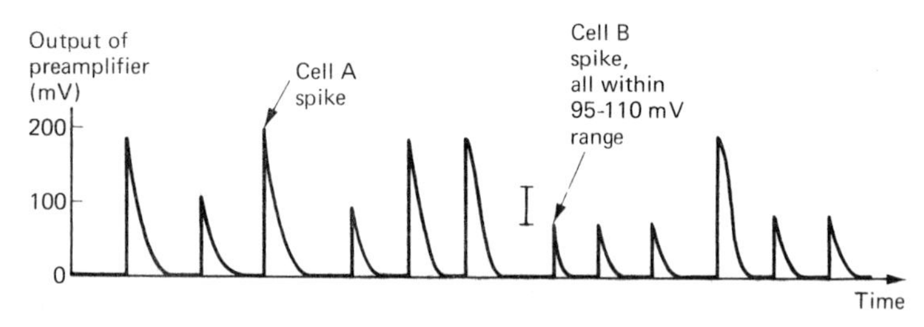


Fig. 5.1

The two alternative approaches are:

(1) connect a preamplifier to a pen recorder, which is simply a voltmeter with a pen replacing the pointer, and a motor drive to move a roll of paper past the pen. It draws a graph of voltage against time. You then sort through the record with a ruler counting the relevant spikes.

(2) build an electronic circuit which outputs a spike every time it receives an input greater than 95 millivolts but less than 110 millivolts. This constitutes an electronic

event-recogniser. Connect the output of this circuit to a counter, zero the counter at the beginning of each test period and write down the counter reading after one minute.

The second method is obviously enormously labour-saving but the chart recorder does have some advantages. For instance if you later decide that you would like to know about the activity of cell A too during the test period you can easily re-analyse the chart record, but with the strongly data-reducing counter method this data is irretrievably lost.

My own compromise is to record the data initially on magnetic tape, a data-preserving storage method which, unlike chart records, enables you to feed the data back into electronic systems afterwards. Then I play back the tapes through a suitable event-recogniser/counter system discarding unwanted data and obtaining the information in which I am interested in a convenient form. If I later decide that some other data are needed I can replay the tape into a different event-recognition system.

## 5.2 ELECTRONIC EVENT-RECOGNISERS

**Comparators** ('triggers' – the well-known Schmitt trigger is the simplest form of comparator)

Comparators are electronic circuits which decide between two outputs depending on the input voltage level (figure 5.2). In effect a comparator is a non-inverting amplifier of such enormous gain that a small input signal will send its output hard over one

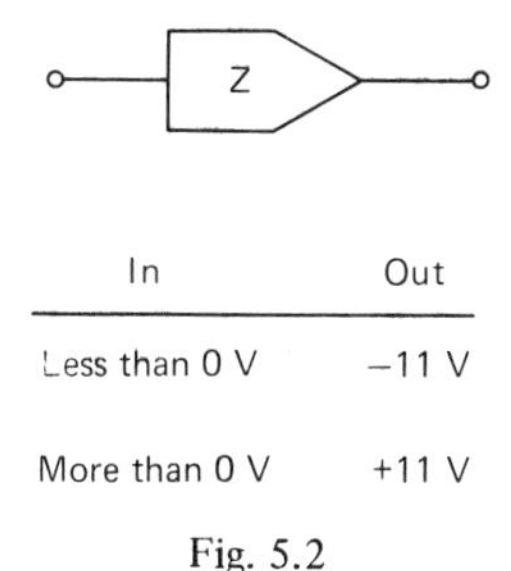


| In | Out |
|---|---|
| Less than 0 V | −11 V |
| More than 0 V | +11 V |

Fig. 5.2

way or the other. Specially designed integrated-circuit comparators are used when very fast operation is needed and these frequently have outputs which go from + 4.5 volts to − 0.5 volts in order to work the digital integrated logic circuits used for computers.

For the typical rather slow biological signal it is possible to use an operational amplifier without negative feedback as a comparator (figure 5.3). The 5 kΩ resistor is to protect the input against excessive currents. Since the amplifier has an input impedance of 400 kΩ the 5 kΩ resistor has a negligible effect on the input voltage to the non-inverting input. Using a -709 series amplifier the gain is 45 000, so all input voltages greater the + 0.25 mV produce the positive full-scale deflection output of + 11 volts, whilst voltages of less than – 0.25 mV produce the maximum negative voltage of – 11 volts.

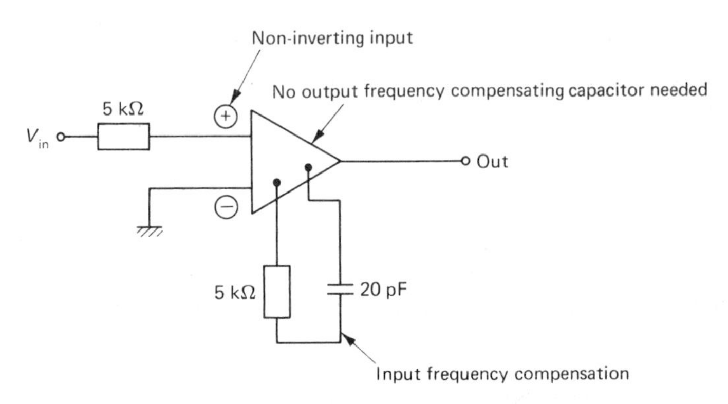


Fig. 5.3

The amplifier gain falls off as the frequency of the input signal increases, even with the small frequency compensation components shown, so the amplitude of the minimum signal required to drive the comparator increases with frequency. To prevent the comparator from jittering between its two output states in doubtful cases (e.g. an input of exactly zero volts) a positive feedback resistor is used (figure 5.4). This means that once in a given state the comparator has a measure of reluctance to change it. If, for instance the output is of + 11 volts, then the current flowing

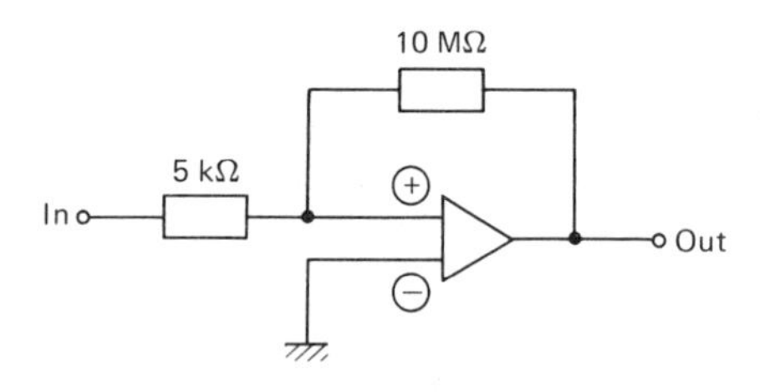


Fig. 5.4

down the 10 MΩ feedback resistor adds + 5 mV to any input signal (figure 5.5). The signal source has to go to – 5 mV to get the comparator input to zero volts and thus change its output to – 11 volts. As soon as this change occurs, however, the feedback resistor adds a negative voltage to the input (figure 5.6). To change the comparator back again the signal source must go to + 5 mV to counteract this (figure 5.7). The last graph shows the actual state of affairs at the amplifier's non-inverting input lead. The change in the feedback resistor's contribution from + 5 to

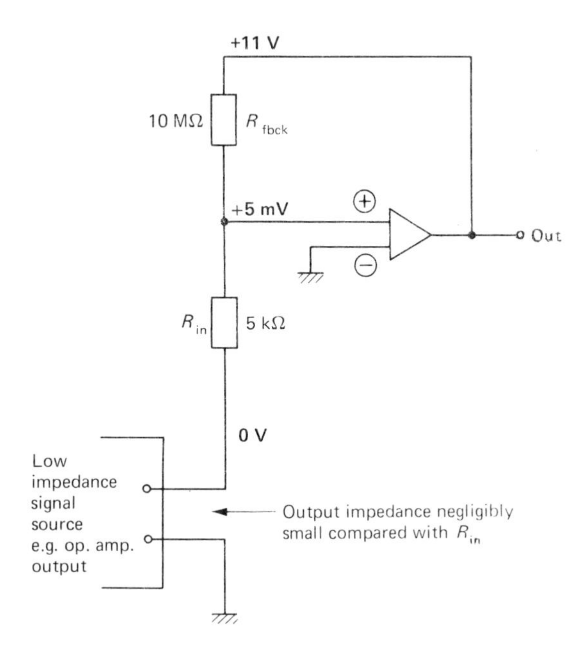
+11 V
10 MΩ
R fbck
+5 mV
Out
R in
5 kΩ
0 V
Low impedance signal source e.g. op. amp. output
Output impedance negligibly small compared with R in

Fig. 5.5

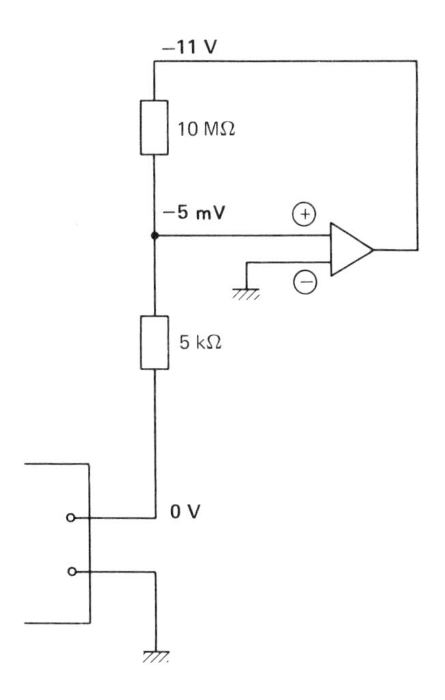
−11 V
10 MΩ
−5 mV
5 kΩ
0 V

Fig. 5.6

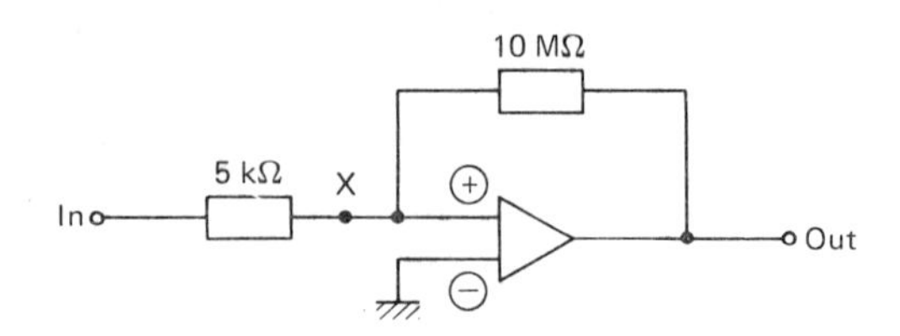
10 MΩ
5 kΩ
X
In
Out

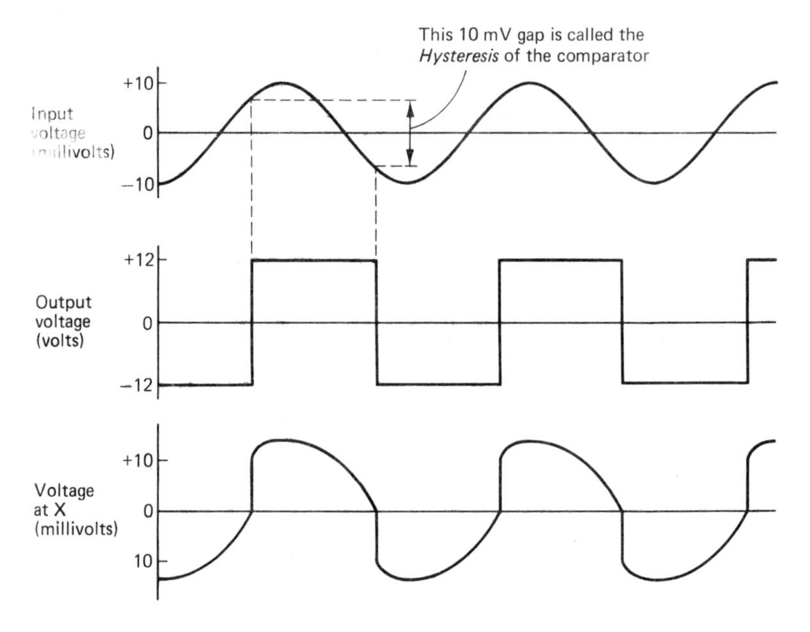
This 10 mV gap is called the
*Hysteresis* of the comparator
+10
0
−10
Input
voltage
(millivolts)
+12
0
−12
Output
voltage
(volts)
+10
0
10
Voltage
at X
(millivolts)

Fig. 5.7

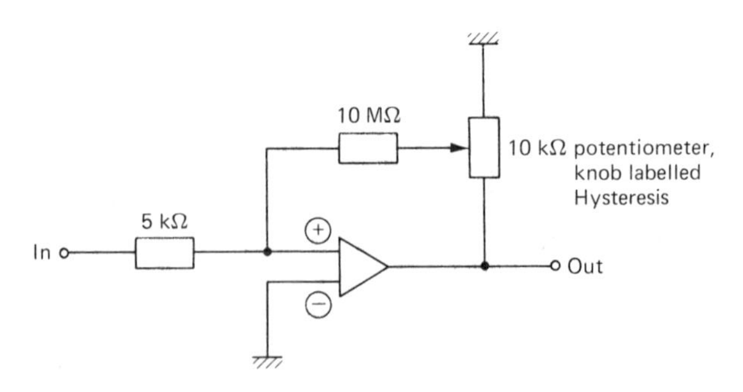
10 MΩ
10 kΩ potentiometer,
knob labelled
Hysteresis
5 kΩ
In
Out

Fig. 5.8

– 5 mV every time the comparator goes over can be seen. Variable hysteresis is best achieved as shown in figure 5.8.

The extent of the hysteresis for any given combination of input resistor and feedback resistor tends in practice to be appreciably greater than that calculated as above by the voltage divider method, which makes several dubious assumptions (for example, that no current flows into or out of the amplifier input).

The voltage at which the comparator goes over can be altered by connecting the inverting input to some voltage other than earth (figure 5.9). With –709 series

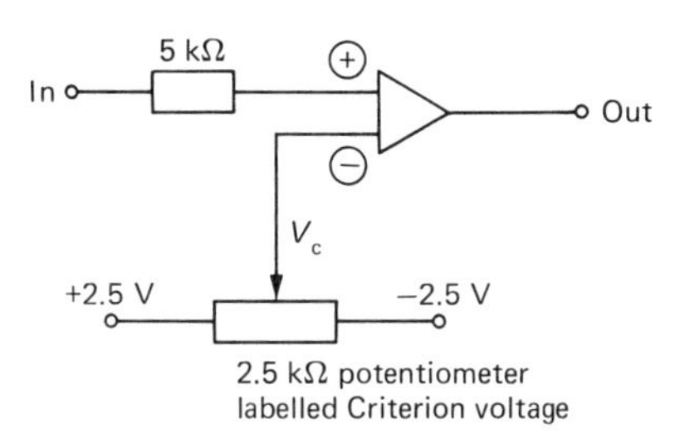


Fig. 5.9

operational amplifiers the voltage between the inverting and the non-inverting input should be kept below five volts, which limits the extent of the adjustment possible by this method.

## Delay Circuits (Monostable Multivibrators)

The second building block of the electronic event-recogniser is the delay circuit (figure 5.10). The output of this circuit stays at zero volts until there is a sudden change in the input voltage. The output then goes to +12 volts, stays there for a

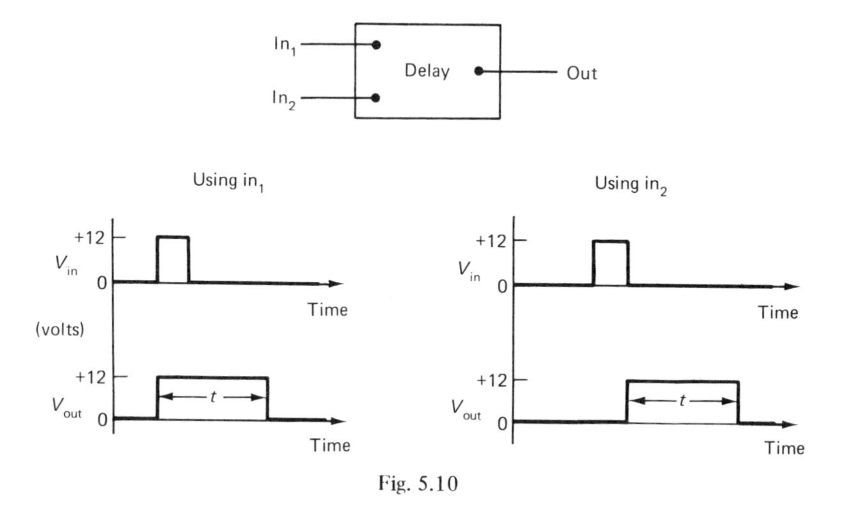


Fig. 5.10

predetermined length of time, then returns to zero volts. There are two inputs, one operated by positive-going voltage changes and the other by negative-going ones. The circuit can be arranged to work on either by connecting the two inputs together (figure 5.11). Input changes during the period in which the output is at + 12 volts have no effect.

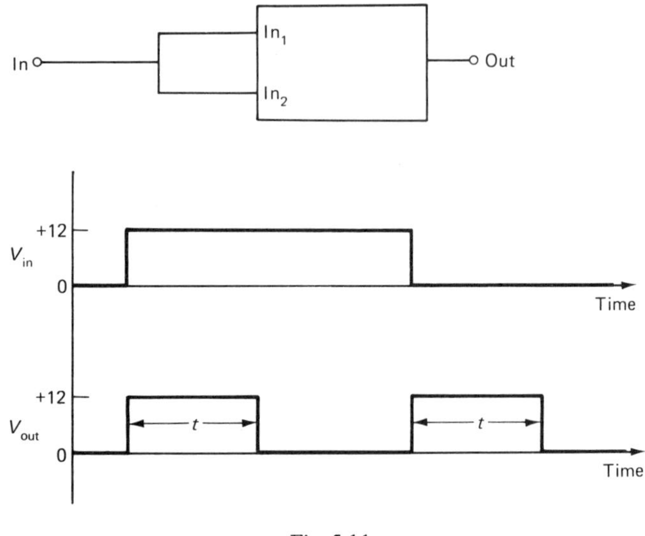


Fig. 5.11

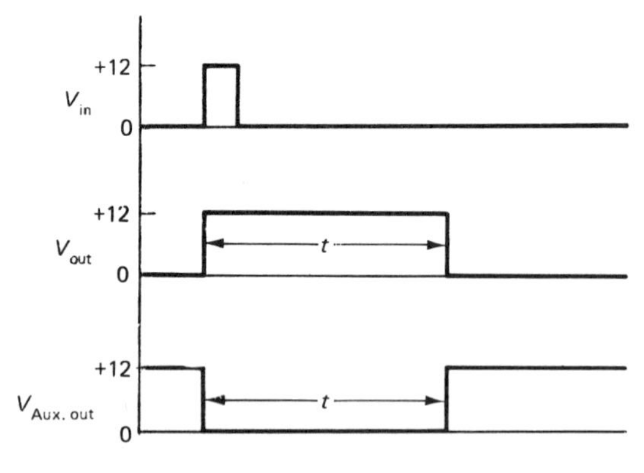


Fig. 5.12

The time $t$ for which the output stays at + 12 volts can be set anywhere between 100 nanoseconds and 1 minute, and can be adjusted by a factor of ten with a potentiometer if a variable delay is needed. There are two more possible connections, an auxiliary output (figure 5.12) which supplies a mirror image of the output pulse, and an input paralyser, which prevents a valid input from operating the circuit (figure 5.13). Versions with outputs which go from 0 to − 12 volts are also available.

This delay circuit, variously called a monostable or a one-shot multivibrator, is available as an integrated circuit, but is very simple to build from discrete components. A description of how it works and instructions for building it are given in chapter 8.

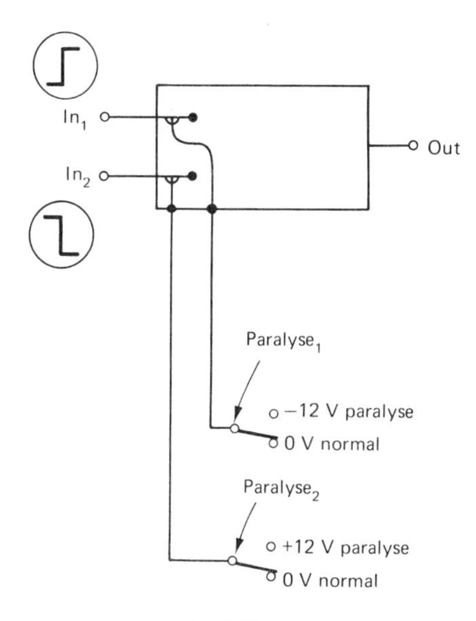


Fig. 5.13

We are now in a position to design an electronic pattern-recogniser which will identify spikes between 95 and 110 mV high (figure 5.14). First the spikes are amplified ten times to make them easier to work with. As the amplifier inverts its input signal the voltage range we are interested in now lies between − 0.95 and − 1.10 volts. The signal is connected to two comparators, one set to go over at − 0.95 volts, and one set for − 1.10 volts. Spikes of less than 95 mV set off neither comparator. Those between 95 and 110 mV set off the upper comparator but not the lower one. The upper comparator's output moving from + 12 to − 12 volts triggers delay 1 via its negative-going pulse input 2. The output of delay 1 goes at once from zero volts to + 12 volts, stays there for 1 ms, then returns to zero volts. This last triggers delay 3, which is used to advance the counter reading by one. So there is a 1 ms delay between the spike setting off the comparator and its being counted.

If the spike is greater than 110 mV it triggers both comparators, first the 0.95 V one, then, a fraction of a millisecond later, the 1.10 V one. Delay 1 is operated as above, but this time Delay 2 is triggered as well by the 1.10 V comparator. As long as delay 2's output is at + 12 volts delay 3 is paralysed and the counter cannot

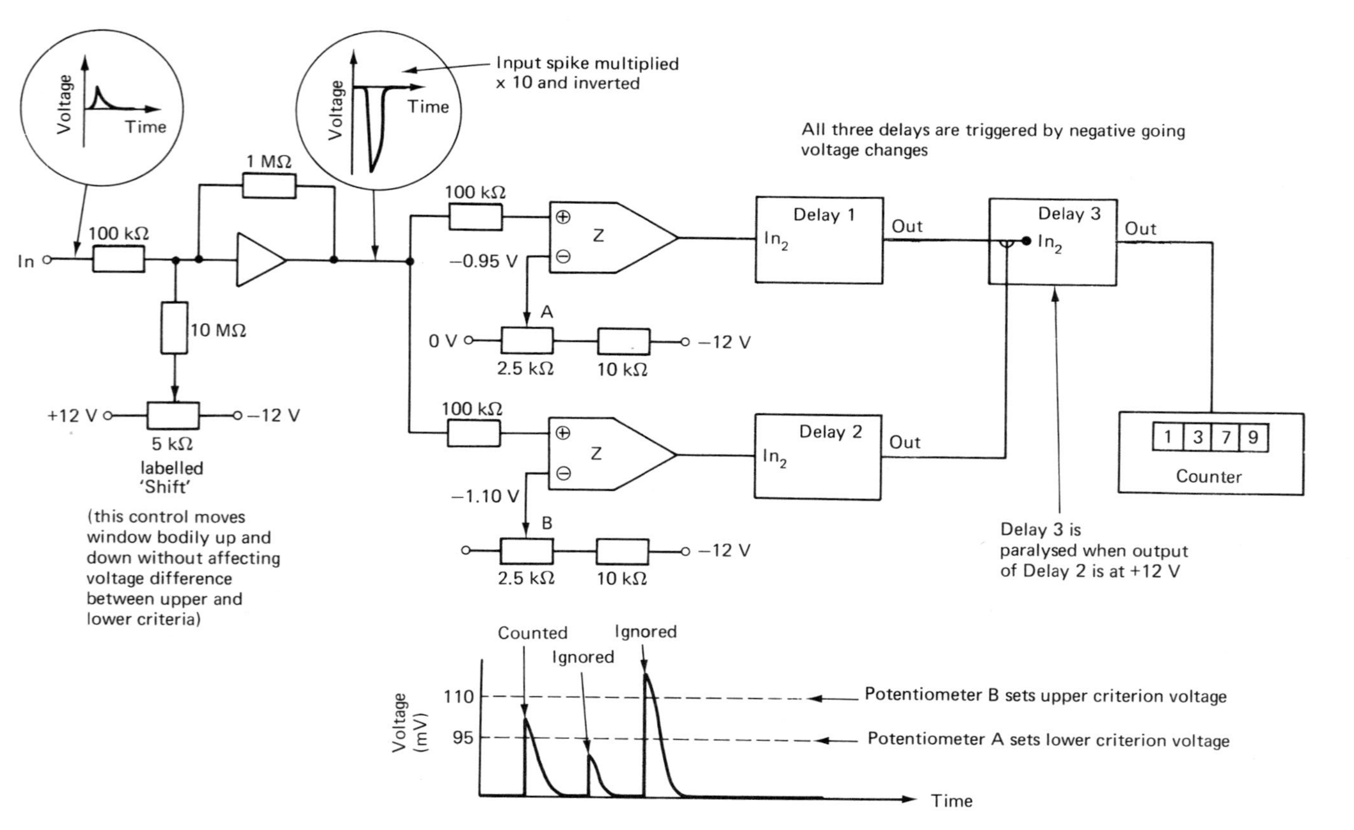
Voltage
Time
Voltage
Time
Input spike multiplied x 10 and inverted
All three delays are triggered by negative going voltage changes
In
100 kΩ
1 MΩ
10 MΩ
+12 V
−12 V
5 kΩ
labelled 'Shift'
(this control moves window bodily up and down without affecting voltage difference between upper and lower criteria)
100 kΩ
Z
−0.95 V
A
0 V
2.5 kΩ
10 kΩ
−12 V
100 kΩ
Z
−1.10 V
B
2.5 kΩ
10 kΩ
−12 V
Delay 1
In2
Out
Delay 2
In2
Out
Delay 3
In2
Out
1 3 7 9
Counter
Delay 3 is paralysed when output of Delay 2 is at +12 V
Counted
Ignored
Ignored
Voltage (mV)
110
95
Potentiometer B sets upper criterion voltage
Potentiometer A sets lower criterion voltage
Time

Fig. 5.14

operate. Since delay 1 is triggered just before delay 2, its output will return to zero volts just before delay 2's does, that is during the time delay 3 is paralysed. Hence the counter ignores pulses which exceeded 110 mV.

This device is called a pulse-height discriminator, and, to use it efficiently, a good oscilloscope display showing the input signal and the position of the upper and lower window edges is needed. This is most simply achieved by using the oscilloscope 'Z-mod' (mod = modulation) facility. This is a socket on the back which turns the trace brightness up when a positive voltage is applied and down for a negative one. Generally + 12 volts will turn the trace as bright as it will go and − 12 volts will turn it off completely. Z-mod inputs are almost always AC-coupled, so the effects of connecting a steady voltage lasts only for 2–3 milliseconds. If the outputs of both comparators are connected to the Z-mod socket via small capacitors to prevent them affecting each other, then there will be a brief blanking out of the trace each time a comparator fires (figure 5.15). The oscilloscope brightness control is set so that the blankings due to negative-going comparator triggerings show clearly, but the subsequent brightenings when the comparator returns to + 11 volts do not show. (Use a diode if this proves troublesome.) Unfortunately the Z-mod affects both beams.

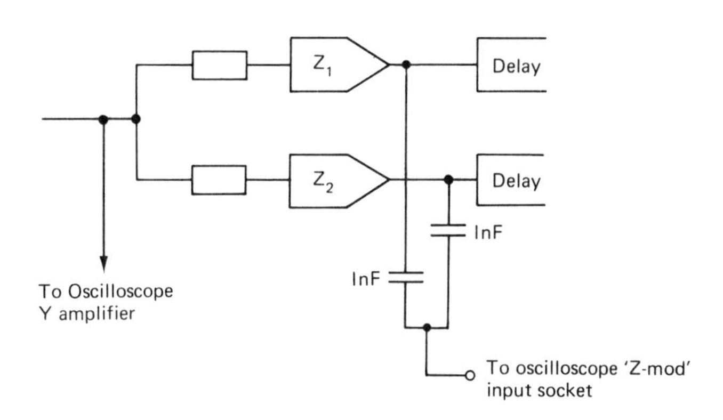


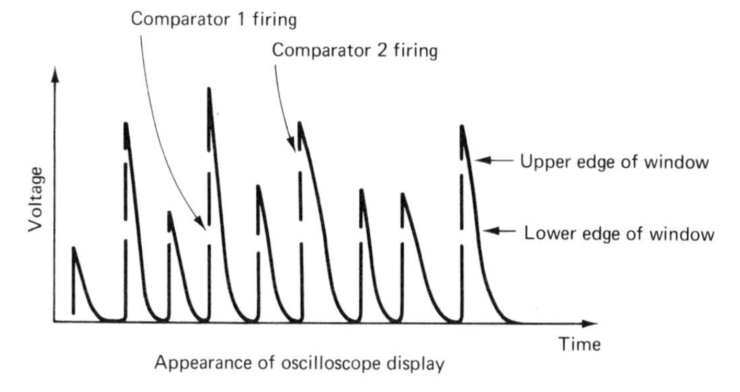


Fig. 5.15

**Integrators**

Integrators are operational amplifiers with a feedback capacitor. This works just like a feedback resistor in stabilising the inverting input at zero volts and preventing any of the current from the input resistor flowing into the amplifier. Figure 5.16 shows the situation for inputs which are steady voltages. The input current all flows into

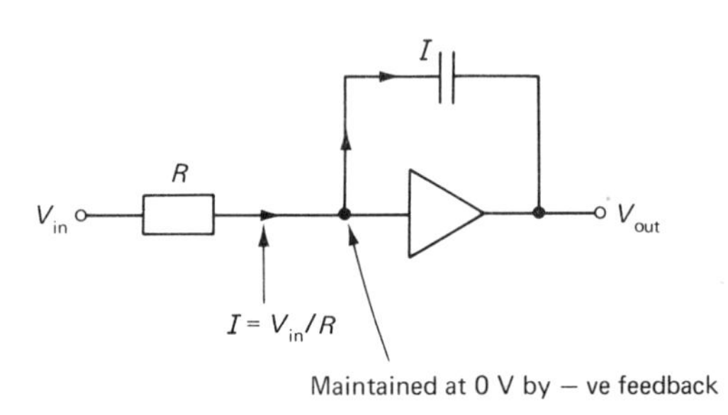


Fig. 5.16

the capacitor, which steadily charges up. As its left-hand end is constrained to zero volts by the effects of feedback, its right-hand end becomes steadily more negative. For any capacitor

$$\frac{\text{Change of volts across capacitor}}{\text{Time taken}} = \frac{\text{Input current}}{\text{Capacitance}}$$

or

$$\frac{dV \text{ volts (or millivolts)}}{dt \text{ seconds (or millisecs)}} = \frac{I \text{ amps (or microamps)}}{C \text{ farads (or microfarads)}}$$

$dV/dt$ is the mathematical symbol meaning the differential of the voltage with respect to time, or in this case the slope of a graph of volts against time. The equation also works for the units shown in brackets; figure 5.17 gives a concrete example.

The larger the steady input voltage, the smaller the resistor and the smaller the capacitor, the steeper is the resulting rise of the output. Negative input voltages produce a positive-going ramp (figure 5.18). For a changing input voltage the output curve is a mathematical function of the input curve and is called the integral (figure 5.19). This means that, at any instant, $V_{out}$ = (total area under curve below line) − (total area under curve above line).

Sometimes integrators are used directly to measure the area under a curve. For instance if a flying insect is being studied in a wind tunnel whose windspeed is varied so that the insect never actually manages to get anywhere, the signal output is the windspeed in the tunnel, corresponding to the flying speed of the insect. If this velocity signal is integrated, the integrator output gives the distance the

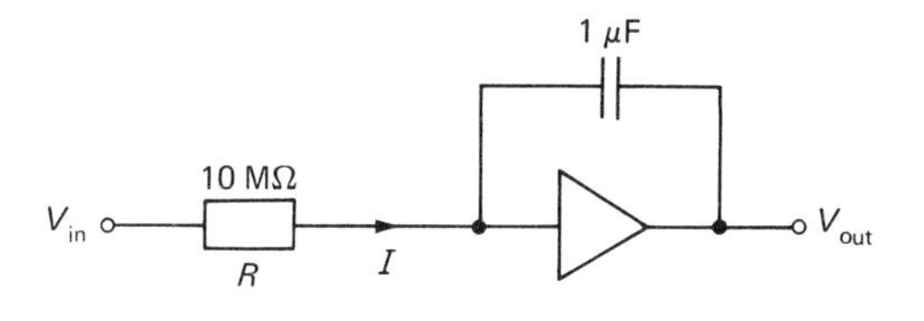


$$\text{for } V_{in} = 1\text{ V};\ I = \frac{V_{in}}{R} = \frac{1}{10^7} = 10^7\text{ A} = 0.1\ \mu\text{A}$$

The rate of charging of the capacitor

$$\frac{dV}{dt} = \frac{0.1\text{ (microamp)}}{1\text{ (microfarad)}} = \begin{array}{l}\text{10 millivolts/millisecond or}\\ \text{10 volts/second}\end{array}$$

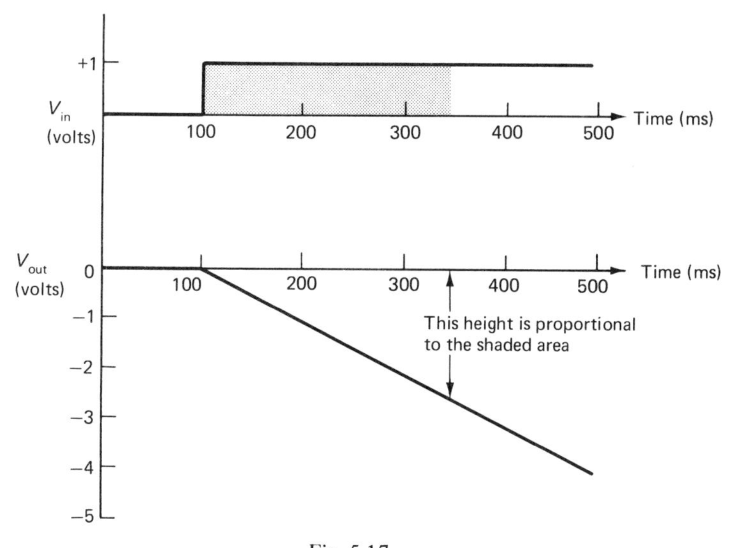


Fig. 5.17

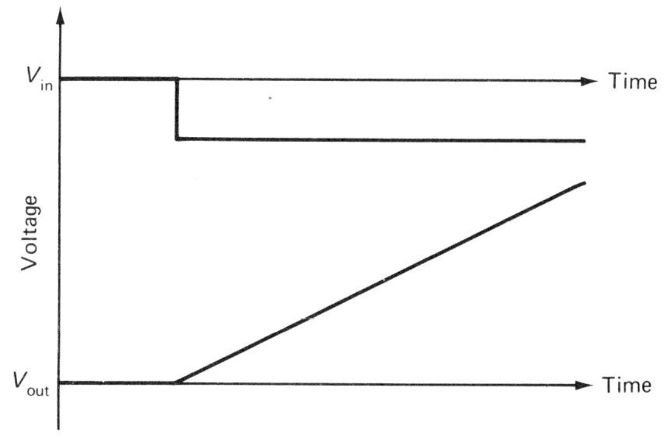


Fig. 5.18

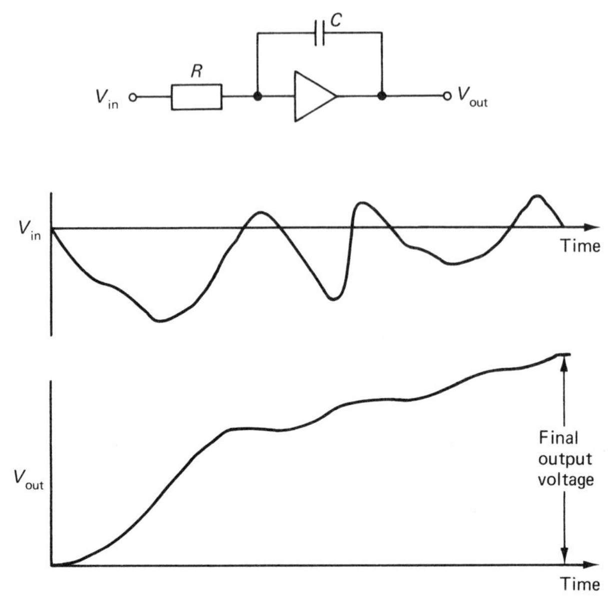


Fig. 5.19

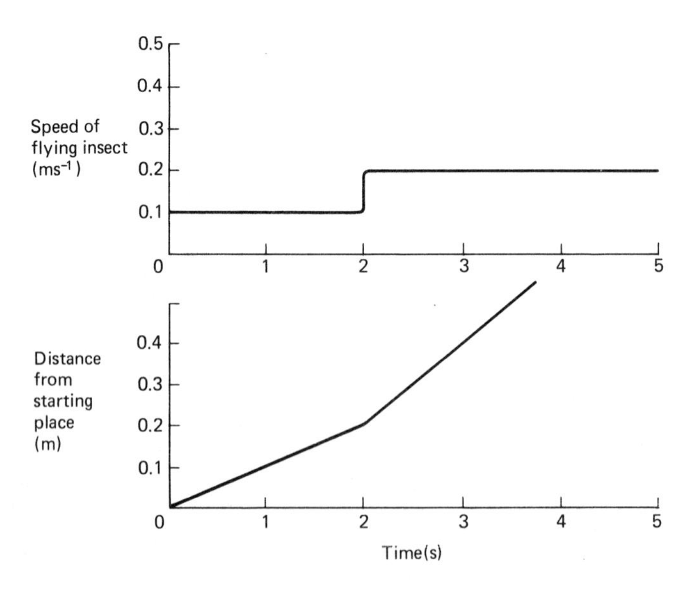


Fig. 5.20

insect would have travelled in the direction of the airflow if it had been flying freely rather than in a treadmill (figure 5.20).

The first problem is that the position graph rather soon disappears off the top of its scale. In physical terms the voltage across the feedback capacitor reaches the power supply voltage and either sticks there, or, with some operational amplifiers, catastrophically continues to rise till it breaks the amplifier. The solution is to discharge the capacitor and start again each time the output reaches a predetermined voltage (figure 5.21). This is achieved in electronic terms by using an FET switch to

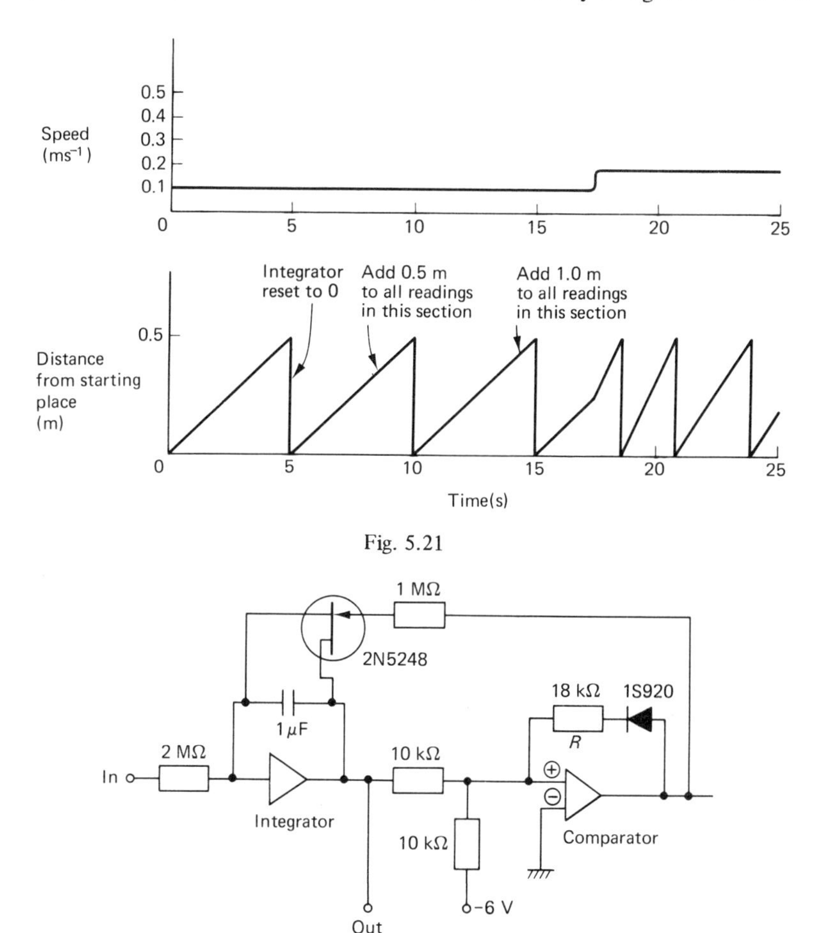


Fig. 5.21

Fig. 5.22

discharge the integrator capacitor whenever a comparator monitoring the integrator output says that it has reached the predetermined voltage (figure 5.22). This circuit starts from the assumption that the insect is flying forwards at varying speeds, and that these speeds are represented by voltages between zero and − 12 volts. So the integrator output trend is always upward, and we only need to reset the integrator when its output exceeds a positive criterion voltage (say + 6 volts), there never being any question of negative output voltages which might require a second comparator with a − 6 volt criterion to stop the comparator going off scale at the negative end too.

The comparator in this circuit adds − 6 volts to the integrator output signal (figure 5.23). (The non-inverting input of a comparator is not, of course, a true adder input maintained at zero volts by negative feedback. But it is known to be at

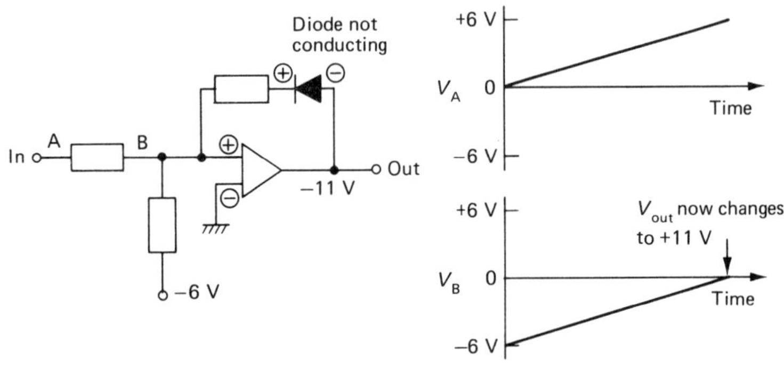


Fig. 5.23

zero volts at the instant the comparator goes over, which is the only time that the adding must be accurate.) The comparator has a diode in its positive feedback loop, so when the comparator output is at − 11 volts (i.e. all input voltages negative of zero volts) there is no hysteresis. When the output is at + 11 volts, however, the diode is conducting and there is a considerable positive feedback current, which must be overcome by the input current before the comparator will return to the − 11 volt state. Given the 6 volt discrepancy introduced between integrator output and comparator input, the comparator remains in its − 11 volt output state, holding the FET switch open, till the integrator reaches + 6 volts.

Then the comparator goes over to the + 11 volts state, closing the FET switch and thus beginning to discharge the capacitor. Because of the hysteresis the switch stays closed till the integrator output reaches zero volts, that is the capacitor is completely discharged. The process can now start again.

The second major problem with integrators is that they accumulate small errors. If the amplifier concerned has a tiny discrepancy in its input circuits a small current (called the offset current) will flow into the capacitor even when the input resistor is grounded, and the output will drift away from zero volts. This drift often accelerates as the output moves further away from zero. It is worth supplying a potentiometer to cancel out such small input currents (figure 5.24). Making this adjustment requires great patience because the potentiometer controls the velocity of the vertical movement of the oscilloscope trace rather than its position. Even with this refinement, and even if you buy very expensive operational amplifiers with extra low offset currents, it is very difficult to make integrators which do not drift perceptibly if operated for periods of more than 1–2 minutes without being reset. For long-term integrators large and very expensive polystyrene dielectric capacitors with very low leakage currents are essential.

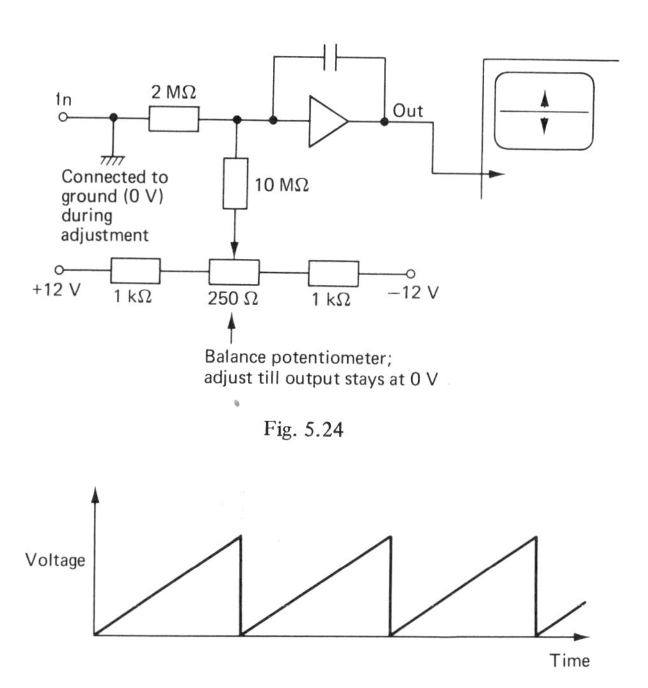


Fig. 5.24

Fig. 5.25

One answer is to make the integrator reset very frequently (small input resistor) and to count the number of resets with a digital counter. The counter reading is proportional to the total area under a graph of input voltage against time (see figure 5.21). If the input to the above resetting integrator circuit is a steady voltage the output is a uniform chain of ramps (figure 5.25). This is the waveform required from an oscilloscope timebase (the repetition rate can be varied by varying the input voltage) and this circuit is frequently used for timebases.

## Staircase Generators

These are integrators with a specially modified input network (figure 5.26). Each time the input moves up to + 12 volts we have the situation shown in figure 5.27. As the input voltage begins its sudden rise to + 12 volts, the 0.2 $\mu$F input capacitor begins to transmit this voltage change to the amplifier inverting input. The amplifier output voltage moves suddenly negative and current flows through the feedback

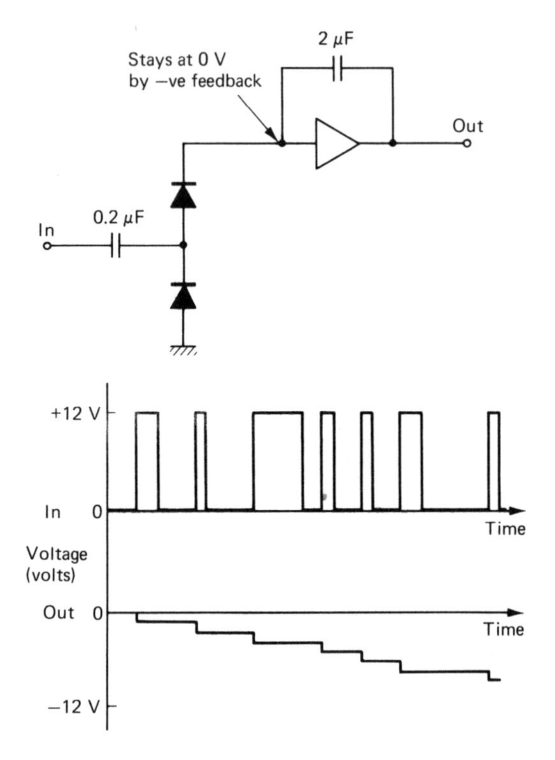


Fig. 5.26

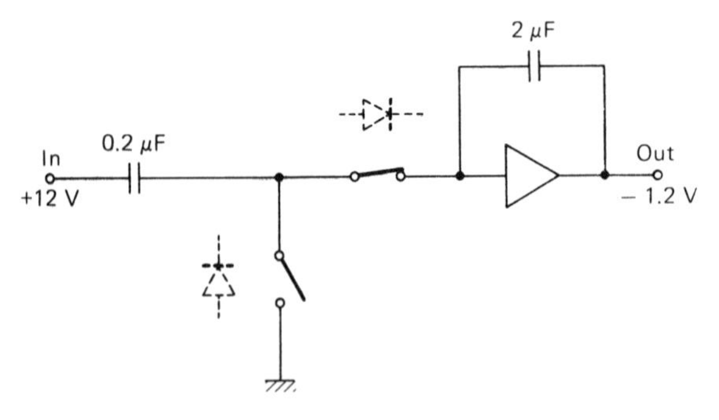


Fig. 5.27

capacitor sufficient to maintain the input at zero volts. When the input voltage reaches a steady + 12 volts the output voltage has reached

$$12 \times \frac{0.2\ \mu F}{2\ \mu F} = (-)\ 1.2 \text{ volts}$$

No further current flow occurs, because the input capacitor transmits only changes in the input voltage and the output capacitor repays these with proportional changes in the output voltage. When the input falls again to zero volts the situation is then that shown in figure 5.28.

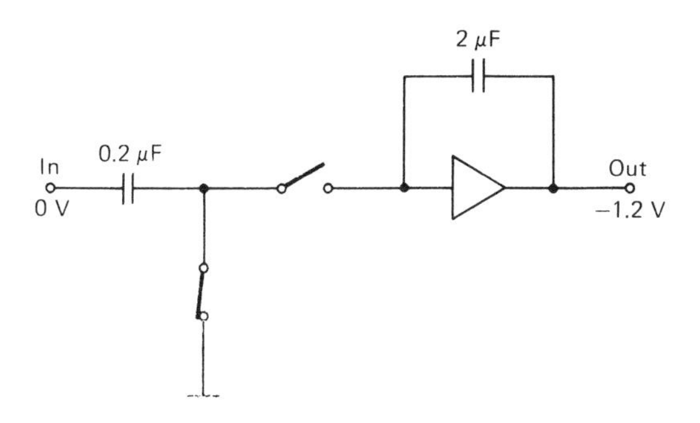


Fig. 5.28

The negative-going voltage change is not transmitted to the amplifier, whose output remains at − 1.2 volts. The input capacitor is discharged, ready to respond to the next 12 volt input pulse. The feedback capacitor has no means of discharging, so the output remains at − 1.2 volts. The next input pulse will require the output to move to − 2.4 volts to transmit enough current through the feedback capacitor to keep the inverting input at zero volts. The size of each step of the resulting staircase curve depends on the ratio of the two capacitors and the height of the input pulses. Provided their length is not very short indeed, it has no effect.

This circuit is very useful for counting every *n*th event in a sequence, a frequent form of data-reduction. The familiar comparator/reset switch circuit generates a single output pulse every time the staircase curve reaches a voltage predetermined by the comparator criterion voltage. A new method of setting the criterion voltage at which the comparator goes over is used in this circuit. This is the most precise of the three methods available (figure 5.29). The circuit using method 3 for a resetting staircase generator is given in figure 5.30. The potentiometer controlling the steady voltage added to the staircase is used to set the number of input pulses required before the comparator goes over to its + 11 volt output state. The comparator's going over triggers a delay circuit whose output closes the FET switch for a time long enough to discharge the integrator capacitor and reset the staircase to zero volts. This is an alternative to the system used for the integrator reset, in which the switch was opened directly by the comparator, and prevented from closing again till the integrator output had reached zero volts by a diode in the feedback loop around the

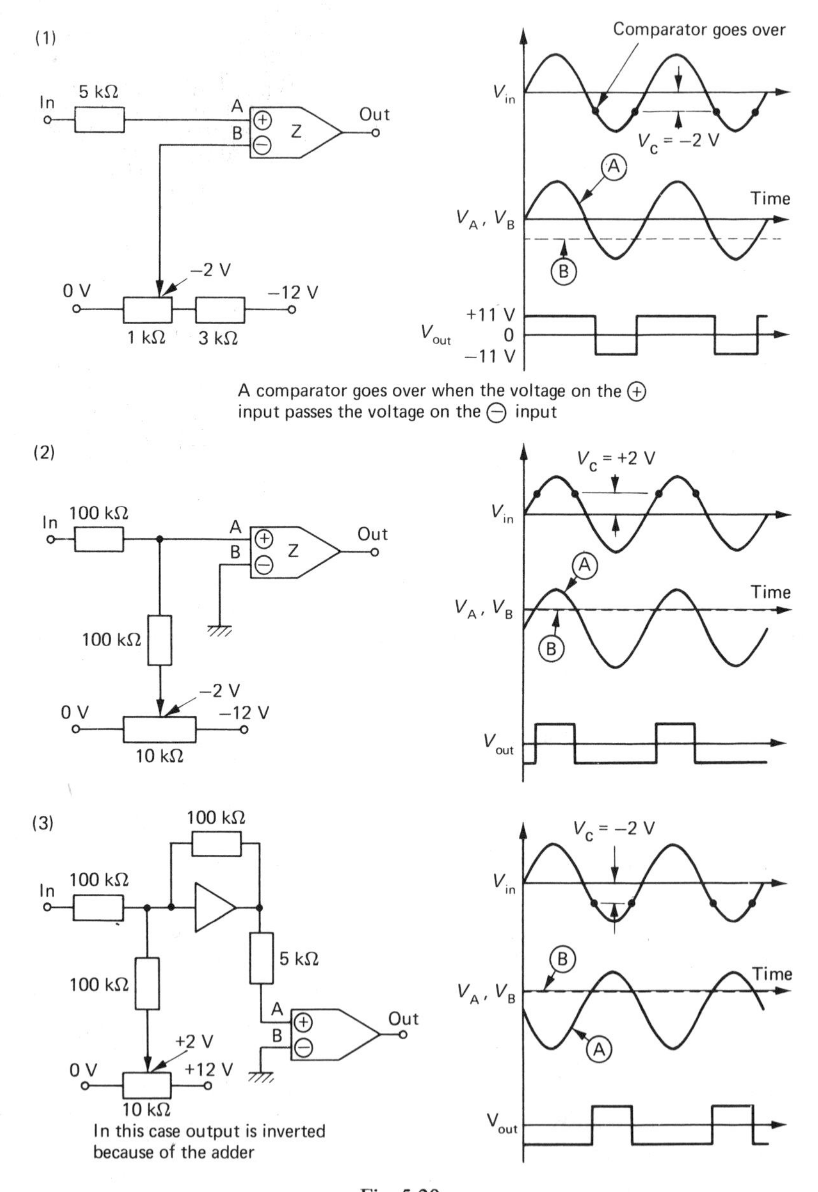
(1)
In
5 kΩ
A
B
Z
Out
0 V
−2 V
−12 V
1 kΩ
3 kΩ
Comparator goes over
$V_{in}$
$V_c$ = −2 V
A
Time
$V_A$, $V_B$
B
+11 V
0
−11 V
$V_{out}$
A comparator goes over when the voltage on the ⊕ input passes the voltage on the ⊖ input
(2)
In
100 kΩ
A
B
Z
Out
100 kΩ
0 V
−2 V
−12 V
10 kΩ
$V_c$ = +2 V
$V_{in}$
A
Time
$V_A$, $V_B$
B
$V_{out}$
(3)
100 kΩ
In
100 kΩ
100 kΩ
5 kΩ
A
B
Out
+2 V
0 V
+12 V
10 kΩ
In this case output is inverted because of the adder
$V_c$ = −2 V
$V_{in}$
B
Time
$V_A$, $V_B$
A
$V_{out}$

Fig. 5.29

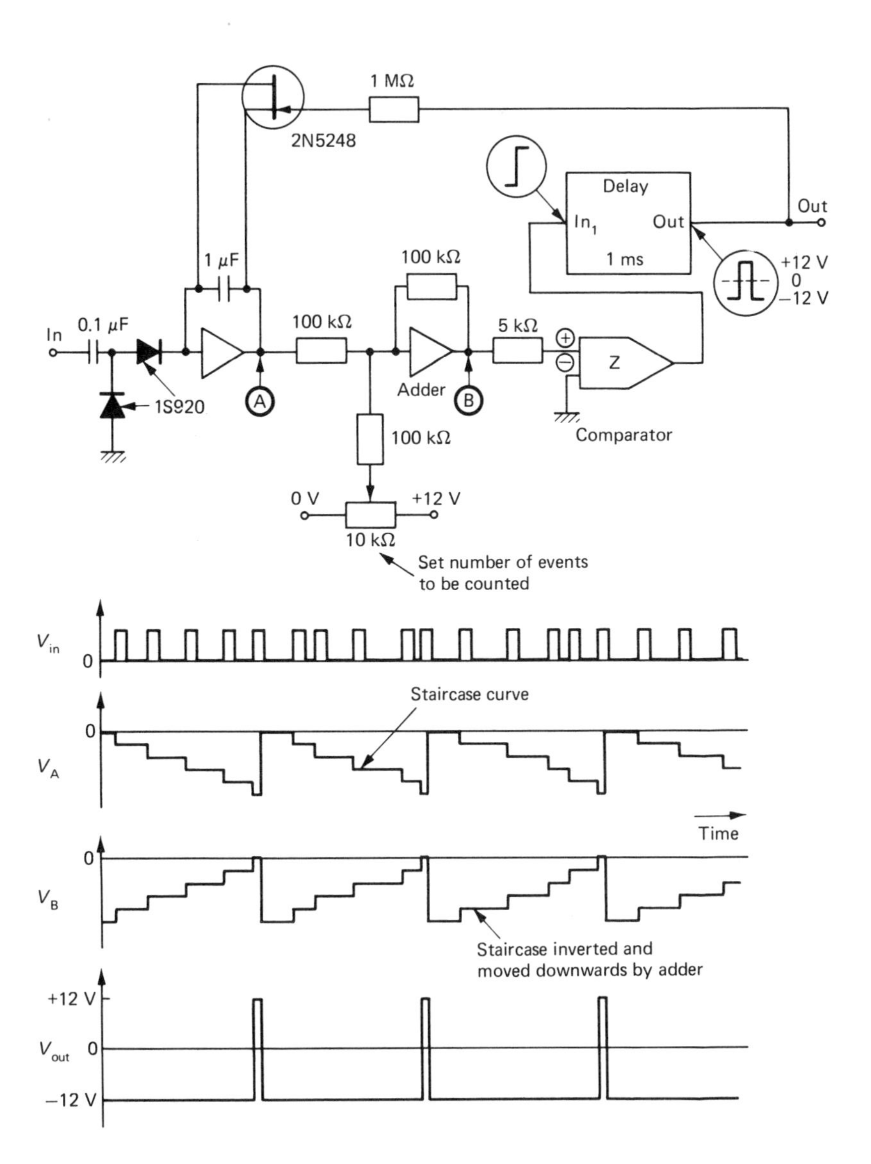
1 MΩ
2N5248
Delay
In1
Out
1 ms
Out
+12 V
0
−12 V
1 μF
100 kΩ
In
0.1 μF
100 kΩ
5 kΩ
Z
1S920
A
Adder
B
Comparator
100 kΩ
0 V
+12 V
10 kΩ
Set number of events to be counted
Vin
0
Staircase curve
0
VA
Time
0
VB
Staircase inverted and moved downwards by adder
+12 V
Vout 0
−12 V

Fig. 5.30

comparator (figure 5.23). The alternative was chosen because in the integrator circuit the value of $R$ (figure 5.23) depends on the criterion voltage setting used. In the integrator this was never altered so $R$ never needed adjusting. This time the criterion is changed every time a different number of input pulses per output pulse is required, so the extra circuitry to hold the switch closed for a fixed time is justified.

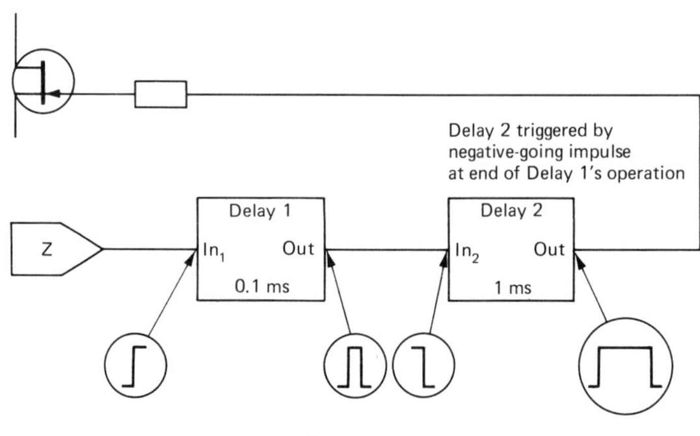


Fig. 5.31

In fact I have found it very difficult to get this delay circuit to operate, because the comparator tends to return to − 11 volts very quickly and stop the delay before it has reached its + 12 volt state. In the end it was necessary to use an extra-brief delay so that the beginning of the discharge of the capacitor did not follow immediately on the comparator going over.

### Sample-and-hold Circuit

The final useful derivative of the integrator is the sample-and-hold circuit. When the switch is in the + 12 volt position this circuit works like an inverting amplifier with unit gain. When the switch is moved to the − 12 position the output sticks at whatever voltage it happens to be at at the instant the switch is moved, and stays there until the switch is changed back again (figure 5.32).

This circuit is often used in conjunction with devices called analogue to digital (A–D) converters, which transform a voltage level into a corresponding binary number suitable for a digital computer. These always take a little time to act, and need their input voltage held steady while they do so. It is also useful if you want to know the maximum voltage reached by a signal source over a given period (figure 5.33). A comparator compares the sample-and-hold circuit output with the inverted input signal. If the input signal is bigger than the sample-and-hold output (i.e. more negative because of the inversion) the comparator goes over and triggers

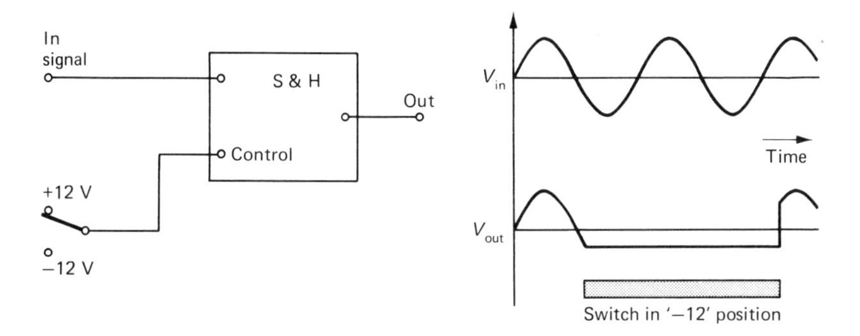


Fig. 5.32

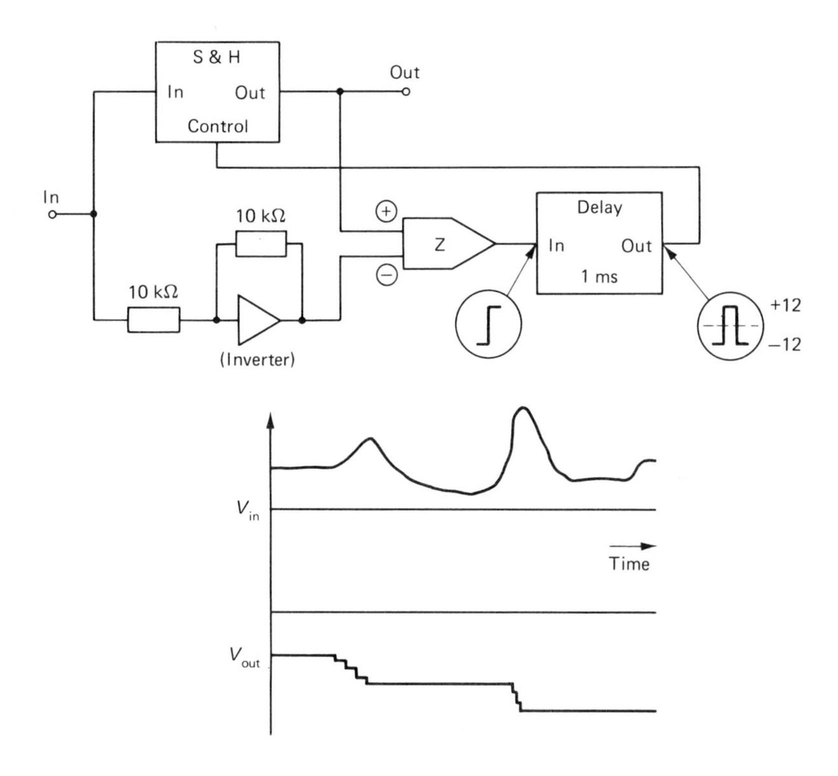


Fig. 5.33

the delay circuit which makes the sample-and-hold take a new sample. If the input signal is smaller than the existing sample no change occurs.

An electronic circuit for the sample-and-hold is given in figure 5.34. With the control input at + 12 volts the FET switch is closed (figure 5.35(a)). The circuit is a unity-gain inverting amplifier. The capacitor limits the amplifier's gain at high frequencies by reducing the effective value of the feedback resistor, but has no other effect. With the control input at − 12 volts and the FET switch open, the situation is as shown in figure 5.35(b). Input and feedback resistors are disconnected from the

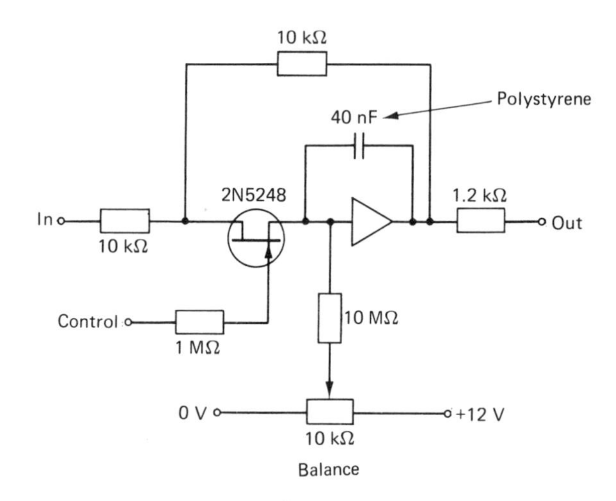


Fig. 5.34

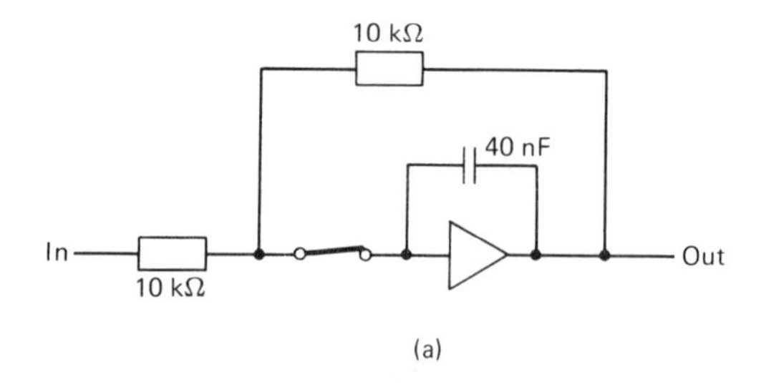


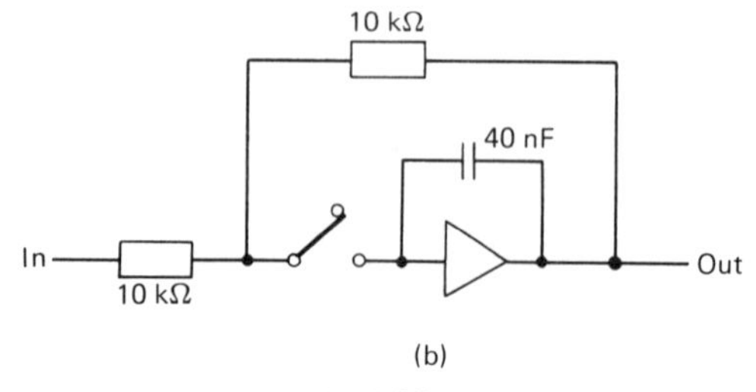


Fig. 5.35

amplifier's input and the only feedback element is now the capacitor. The feedback loop maintains any existing voltage across the capacitor in usual integrator fashion, and so the output remains just where it was when the switch opened.

## 5.3 COUNTERS

The electronic pattern-recogniser represents a particular biological event by a square 12 volt pulse. To study the frequency of the event and the ways in which this frequency is altered by differing experimental situations we need counters, and means of recording the counter readings at specific times.

### Electromagnetic Counters

The simplest, cheapest system is the electromagnetic counter (figure 5.36). This can count up to 50 times per second in its most sophisticated versions, and the extremely cheap 4 number, 10 counts per second versions like that shown below are available on the surplus market. (They are used for adding up telephone bills).

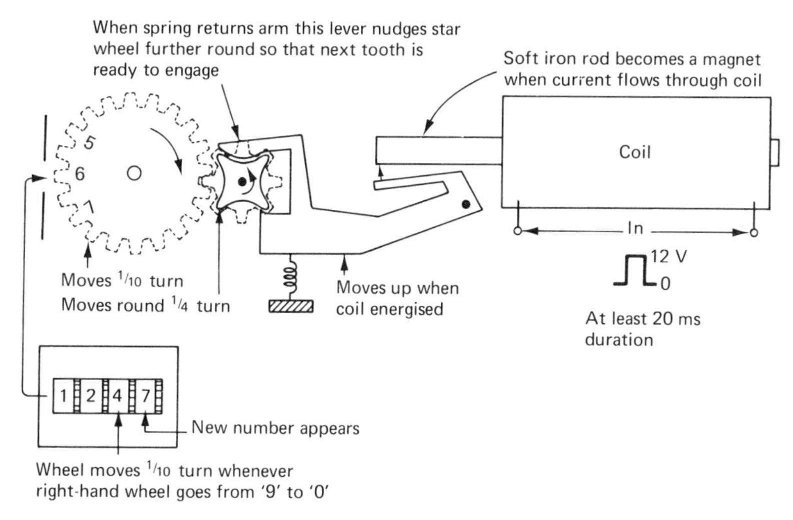


Fig. 5.36

The counter coils have a resistance of around 500 Ω and draw too much current to be worked directly by a delay circuit. It is necessary to interpose a transistor switch. Transistors are the simplest semiconductor devices for making a small current control a big one (figure 5.37). If there is no input current there is no output current either – the transistor is 'hard off', and $V_{out}$ = + 12 volts. As the input current increases, the output current grows in proportion, the voltage across the 2 kΩ resistor increases and $V_{out}$, the voltage on its lower end, is pushed down from + 12

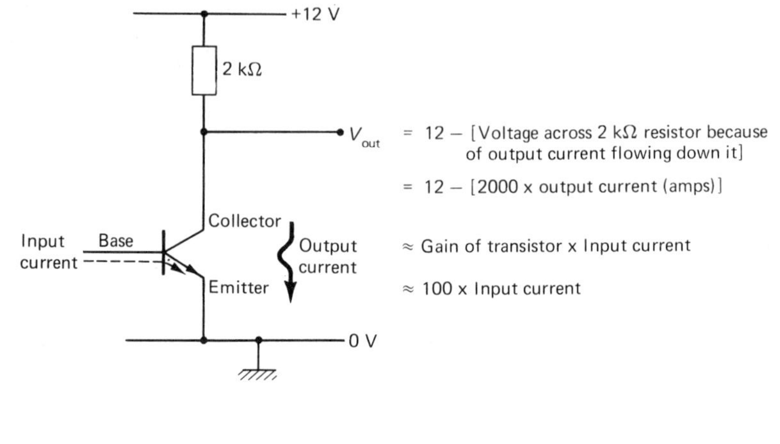


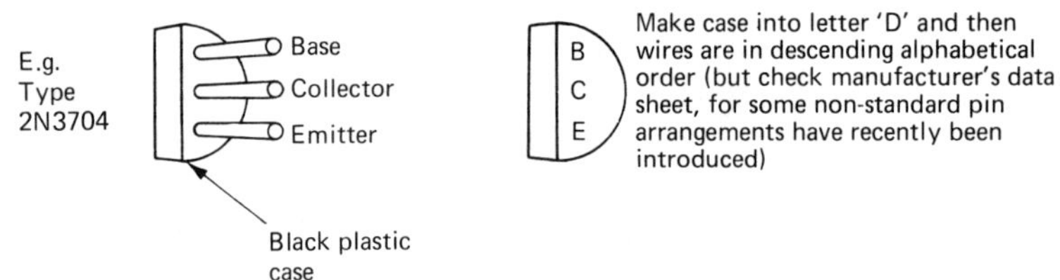


Fig. 5.37

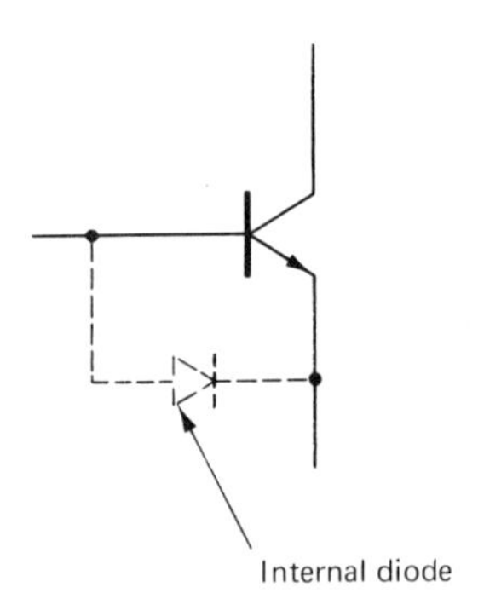


Fig. 5.38

volts towards zero volts. When it gets there, and cannot, of course, go any further (further increase of input current produces no further increase of output current) the transistor is said to be *hard on*. It is normal, when turning on a transistor switch, to allow a safety margin and supply three or four times the minimum necessary input current. Transistors behave as though they had a diode connecting their bases and emitters (figure 5.38). As long as a positive input current is flowing the diode conducts and the base voltage is equal to the emitter voltage. If an attempt is made to draw a current out of the transistor base the internal diode prevents any current actually flowing and the base can take up a negative voltage with respect to the emitter. Most modern transistors are made of silicon and are broken if the base is taken more than – 5 volts from the emitter.

Transistors come in two main types: the one shown in figure 5.37 uses a positive input current to control a positive output current and is called an NPN type. The other sort have negative input and output currents and are called PNP types. (The letters refer the distribution of charge-carrying trace substances in the silicon blocks from which transistors are made.) (figure 5.39.)

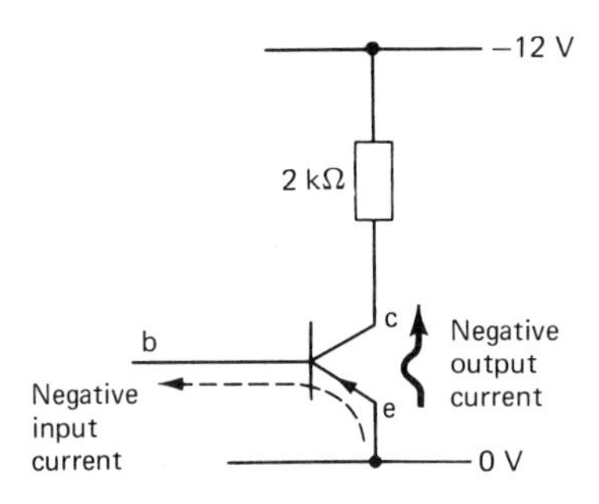


PNP transistors have their internal diode the other way round (hence the different symbol)

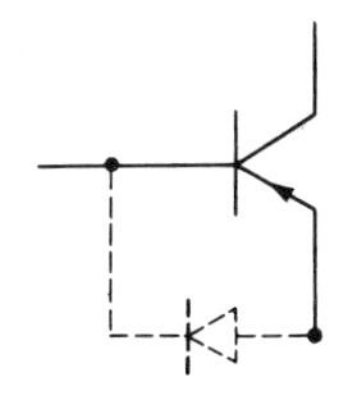

Fig. 5.39

To work a counter, a one transistor switch often suffices (figure 5.40). When the delay-circuit output is at zero volts no input current flows into the transistor, so no output current flows through the counter. When the delay-circuit output rises to + 12 volts a current flows through $R$. As the diagram shows, this current must be 1.2 mA if the transistor is to go hard on. As base volts = emitter volts = 0, the required resistance $R$ is given by

$$R = \frac{V}{I} = \frac{12}{1.2 \times 10^{-3}} = 10\,000\ \Omega$$

Coils tend to give trouble when you switch off currents passing through them. The resulting change in magnetic field induces currents in the coil causing a big overshoot which could easily damage the transistor (figure 5.41). Hence the diode

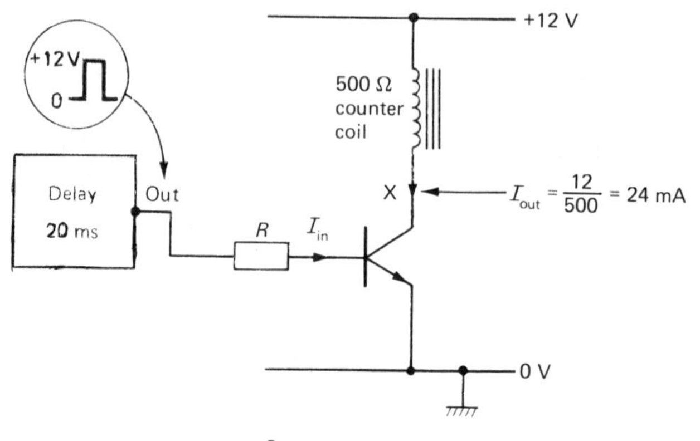


Fig. 5.40

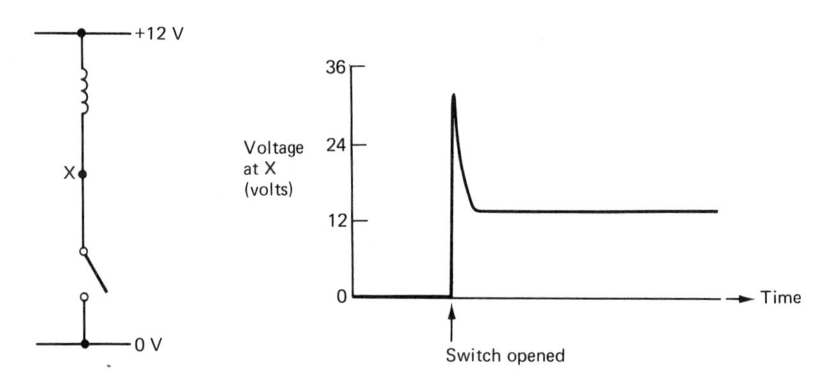


Fig. 5.41

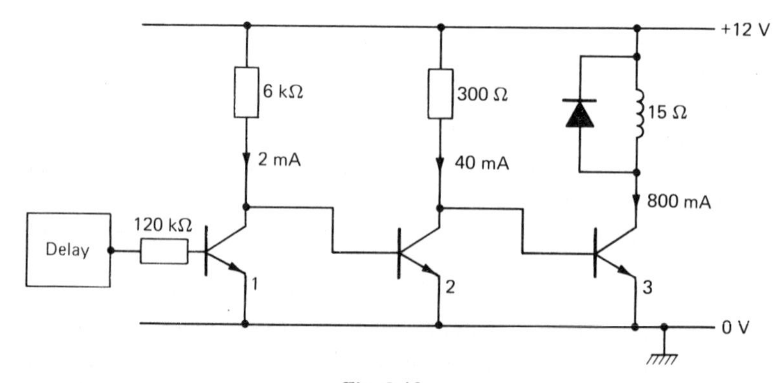


Fig. 5.42

around the coil, which start conducting only if point X exceeds 12 volts and thus shorts out overshoots without affecting the normal working of the circuit.

The transistor used in this circuit is a type number 2N3704. Different transistor types differ in their ability to cope with very high frequency signals, the maximum voltage they can stand between collector and emitter (30 volts in this case), the maximum current that can pass between collector and emitter (0.8 amps in this case) and the maximum power they can dissipate (0.36 watts in this case). Manufacturers' data sheets tabulate these data. In the case of switching circuits the power dissipation in the transistor (volts across transistor times current through it) is small, because when the transistor is off there is 12 volts across it but no current flowing, and when it is on 24 mA flows, but there is only a fraction of a volt between collector and emitter. The limiting factor is the maximum rated current of 0.8 amps, which means that the smallest coil resistance which this particular transistor could cope with is 15 Ω. In such a case the switch would need extra transistors because the input current of the transistor driving the coil would need to be around 40 mA (figure 5.42). If transistor 2 is off, the 300 Ω resistor connected to the + 12 volt rail supplies the necessary input current to transistor 3 and current flows through the counter coil (figure 5.43(a)). If the transistor 2 is on, point X is

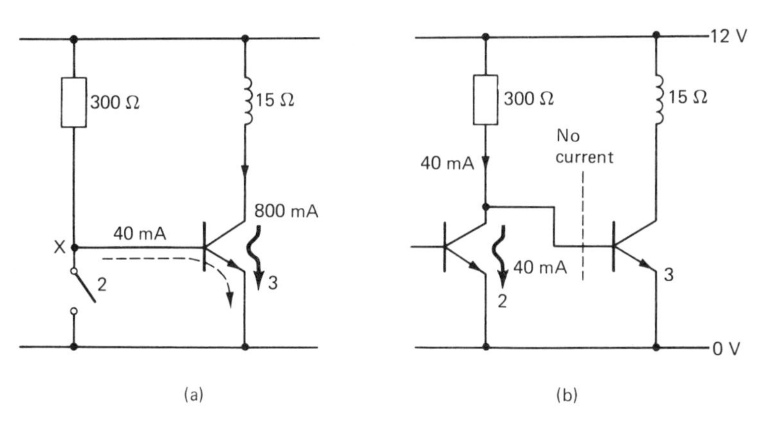


Fig. 5.43

connected to ground and the 40 mA all goes through transistor 2 to ground, instead of into the base of transistor 3 (figure 5.43(b)). So transistor 3 is off, and no current flows through the counter coil. In exactly the same manner if transistor 1 is on it steals all of transistor 2's base current and turns it off. If the delay output is + 12 volts, current flows through the counter coil; if the delay output is zero volts, no current flows (figure 5.44(a), (b)). In a chain like this if any transistor is on, its neighbour to the right must be off and vice versa.

Designing a transistor switch is very easy because the safety factor is built in by assuming that the transistors have a very low gain. The assumption used to calculate the values of the resistors in the above circuit was that the gain of each

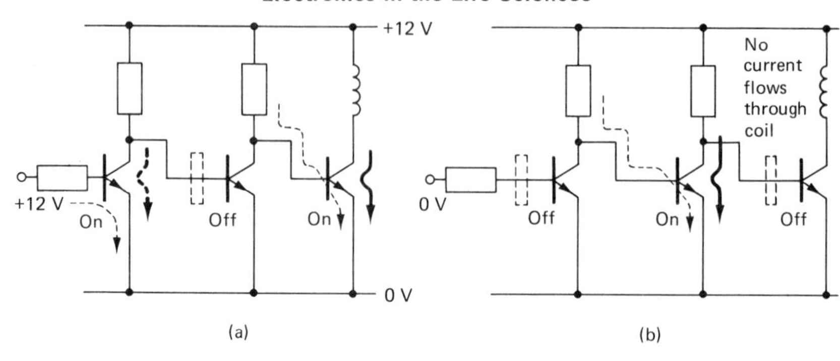


Fig. 5.44

transistor was twenty times. Each transistor's input current was arranged to be one twentieth of its output current by making each resistor twenty times as large as its neighbour to the right. The coil had a resistance of 15 ohms, so the value of transistor 2's collector resistor was chosen to be 15 x 20, = 300 ohms. Similarly transistor 1's collector resistor was 300 x 20, = 6000 ohms, and its input resistor was 6000 x 20, = 120 000 ohms.

### The Printing Counter

The printing counter is a development of the electromagnetic counter in which the numbers on the counter wheels are made of printers' type, and the output of the counter can be recorded automatically on a roll of paper by using an electromagnet ('the *print* coil') to press the paper against the numbers. The most useful versions have two or three independent counters which all print side-by-side on the same roll of paper. These can be made each to count a different category of event, and the answers are all printed out, say once every five seconds. If the events are rare it may be better to use the counter to count seconds and to work the print coil when the event happens. In this case the record shows the time at which each event occurs. Versions of the delay circuit which produce a steady chain of pulses at fixed intervals, called multivibrators, are discussed in chapter 8. These can be used as timers for printing counters for intervals up to about ten seconds. For longer times it is best to use a process timer, a small motor-driven electric clock which closes some contacts when it reaches a time set by a dial pointer. These are described in detail in chapter 7.

The major snag with printing counters is that they must not advance their count while printing, which takes about a tenth of second. This can be achieved by paralysing the delay circuit driving the count coil with the output of that driving the print coil, for when the paralyse terminal of the count delay is held at + 12 volts, it cannot respond to any input pulse (figure 5.45). However there still remains the problem of what happens if the print command arrives while the counter is in the act of advancing its wheels. The only solution is to do nothing about instructions to print till the next event to be counted arrives, count that event, then use the count delay circuit's downward impulse at the end of the count to trigger the print delay and initiate printing.

This involves yet another member of the delay circuit family, the bistable, or memory element (figure 5.46). As soon as this circuit receives an input in the form

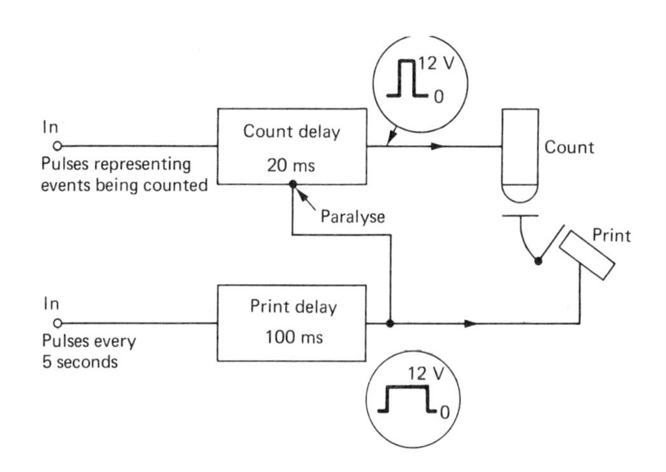


Fig. 5.45

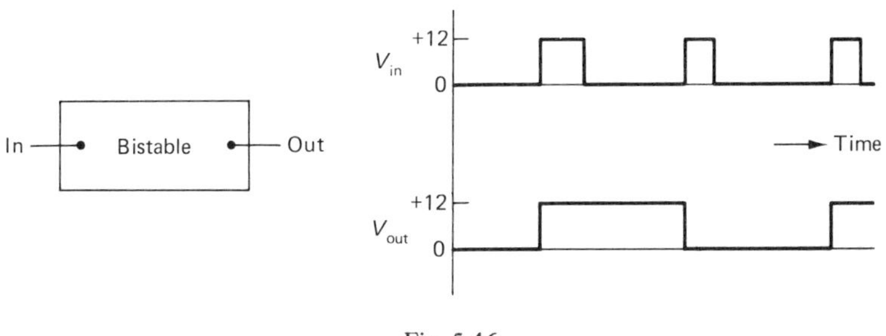


Fig. 5.46

of a positive-going pulse its output moves from zero to + 12 volts and stays there indefinitely, unless another positive-going input pulse arrives, when it returns to zero volts. It works, in fact, just like a retractable ballpoint pen. In this case it is used to inhibit the print delay circuit unless a print command arrives (figure 5.47). The print command pulse pushes the print bistable to its zero volt state, which stops the print delay circuit being paralysed but does not make it go off. Nothing else happens till the next count pulse arrives. The count delay operates, and the downward pulse at the end of its operation triggers the print delay. This works the print coil and also paralyses the counter. At the end of the print delay circuit's operation a positive-going pulse from its auxiliary output is used to push the bistable

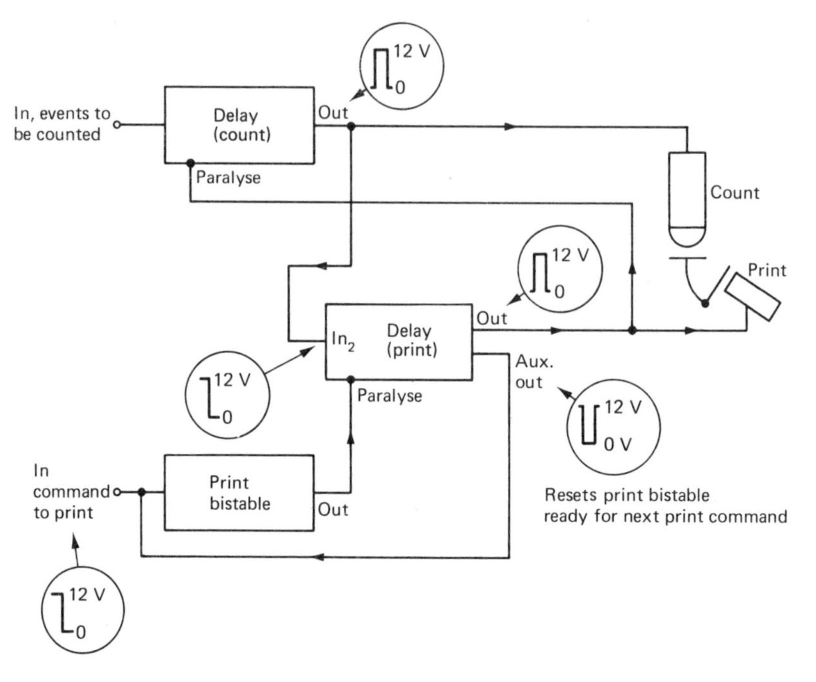


Fig. 5.47

memory back to its + 12 volt state, reimposing the paralysis of the print delay circuit. Information arriving during the printing operation is lost.

**Electronic counters**

An electronic counter is necessary for counting events occurring faster than fifty per second. This displays its counts as glowing numbers formed by suitably shaped electrodes in a vacuum tube, and can cope with up to eighteen million events per second. The actual counting is done electronically by a series of integrated-circuit bistables (see chapter 8) which store the number of events to date in the form of a binary number whose '1' 's are circuits with their outputs in the + 5 volt state and whose '0' 's are circuits with their outputs in the zero volt state (figure 5.48). When an appropriate command signal is received a further set of bistable circuits associated with each display tube takes on values corresponding exactly to those of the counter bistables without interrupting the counting. The tubes light up the appropriate decimal digits corresponding to the binary numbers represented by this second set of bistables, and the display reading remains unchanged until another instruction to sample the counter-circuit outputs is received, despite the fact that the actual internal counter is still acquiring data.

In order to make the translation between binary and decimal numbers simpler, electronic counters generally use a rather awkward counting system called binary

Bistable units of a variety with inputs needing negative-going inputs are convenient:

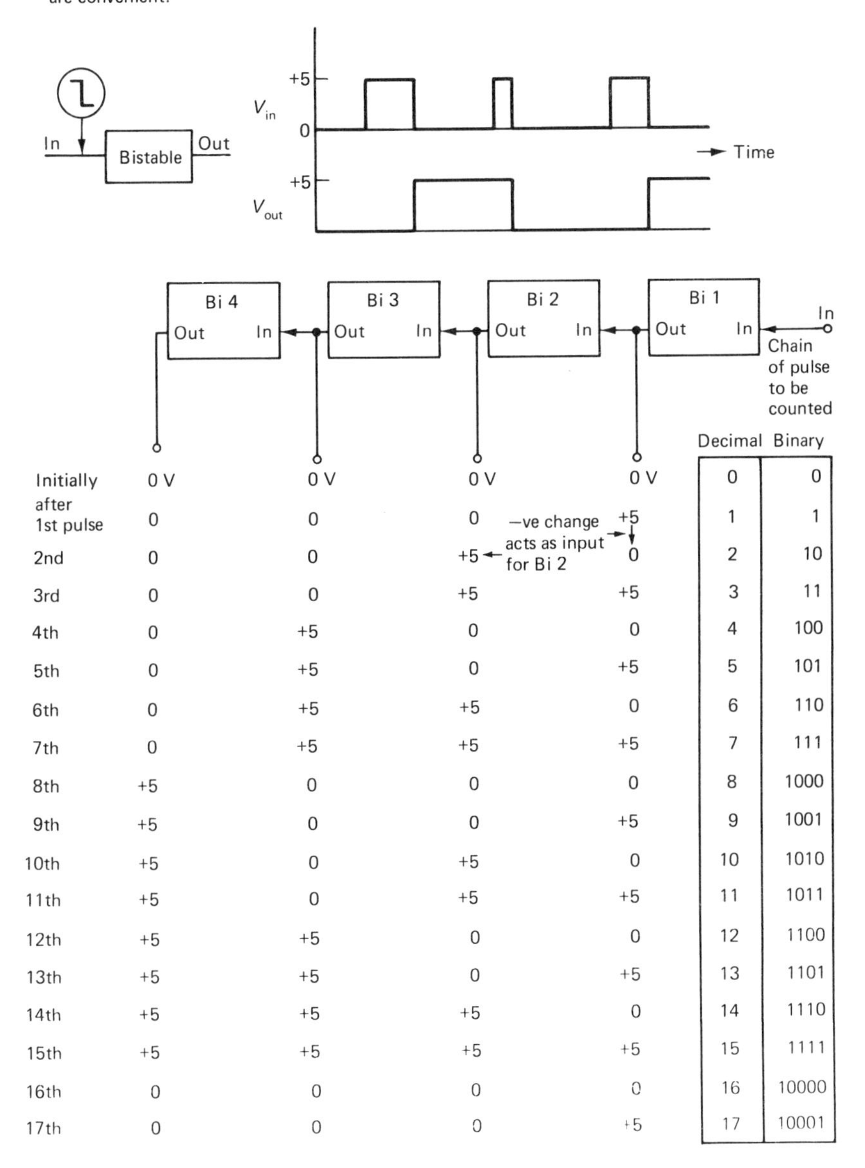


| | Bi 4 | Bi 3 | Bi 2 | Bi 1 | Decimal | Binary |
|---|---|---|---|---|---|---|
| Initially | 0 V | 0 V | 0 V | 0 V | 0 | 0 |
| after 1st pulse | 0 | 0 | 0 | +5 | 1 | 1 |
| 2nd | 0 | 0 | +5 | 0 | 2 | 10 |
| 3rd | 0 | 0 | +5 | +5 | 3 | 11 |
| 4th | 0 | +5 | 0 | 0 | 4 | 100 |
| 5th | 0 | +5 | 0 | +5 | 5 | 101 |
| 6th | 0 | +5 | +5 | 0 | 6 | 110 |
| 7th | 0 | +5 | +5 | +5 | 7 | 111 |
| 8th | +5 | 0 | 0 | 0 | 8 | 1000 |
| 9th | +5 | 0 | 0 | +5 | 9 | 1001 |
| 10th | +5 | 0 | +5 | 0 | 10 | 1010 |
| 11th | +5 | 0 | +5 | +5 | 11 | 1011 |
| 12th | +5 | +5 | 0 | 0 | 12 | 1100 |
| 13th | +5 | +5 | 0 | +5 | 13 | 1101 |
| 14th | +5 | +5 | +5 | 0 | 14 | 1110 |
| 15th | +5 | +5 | +5 | +5 | 15 | 1111 |
| 16th | 0 | 0 | 0 | 0 | 16 | 10000 |
| 17th | 0 | 0 | 0 | +5 | 17 | 10001 |

Fig. 5.48

coded decimal (BCD) (table 5.1). BCD misses out the binary numbers from 1010 (10) to 1111 (15) and so on, in order that each decade of decimal numbers can have its own independent set of four bistables, and just needs a single input pulse each time the previous decade goes from 9 (1001) to 0 (0) to operate it.

Table 5.1

| Decimal | Binary | BCD |
|---|---|---|
| 1 | 1 | 1 |
| 2 | 10 | 10 |
| 3 | 11 | 11 |
| 4 | 100 | 100 |
| 5 | 101 | 101 |
| 6 | 110 | 110 |
| 7 | 111 | 111 |
| 8 | 1000 | 1000 |
| 9 | 1001 | 1001 |
| 10 | 1010 | 10000 |
| 11 | 1011 | 10001 |
| 12 | 1100 | 10010 |
| 13 | 1101 | 10011 |
| 14 | 1110 | 10100 |
| 15 | 1111 | 10101 |
| 16 | 10000 | 10110 |

Electronic counters are unpopular amongst biologists because they often have complicated and unergonomic controls enabling them to be used either as frequency meters or timers over a wide range of frequencies. Moreover they are very prone to interference just because they can operate so fast and tend to add on the odd million to your data everytime they pick up a stray radio signal. The root of this trouble is that the counter comes packaged with electronic clocks, input circuits and automatic reset systems designed to suit electronic engineers' purposes rather than biologists! A good strategy if you need a fast counter is to buy your counters by

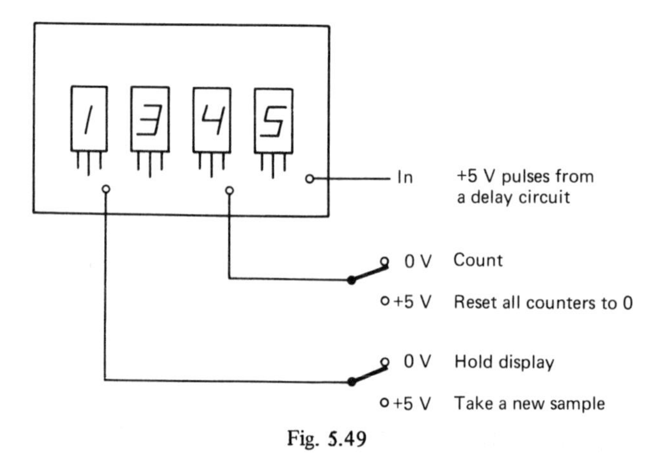


Fig. 5.49

the decade in the form of a number indicator, a printed circuit and three integrated circuits; then connect the decades to each other and to appropriate power supplies, and drive the reset and display controls from external timing circuits (figure 5.49). The resulting device fits into data acquisition systems in exactly the same sort of way as the electromagnetic counters described earlier in the chapter, and costs about one-tenth as much as a commercial electronic counter.

### Rate Meter

These give a meter reading (or output voltage) proportional to the rate of input pulses and can sometimes be used instead of electronic counters to deal with events occurring between 50 and 1000 times per second. They are a simple form of staircase generator in which the integrator capacitor's charge is allowed to leak slowly away through a big resistor (figure 5.50). The output voltage $V_{out}$ is directly proportional to the

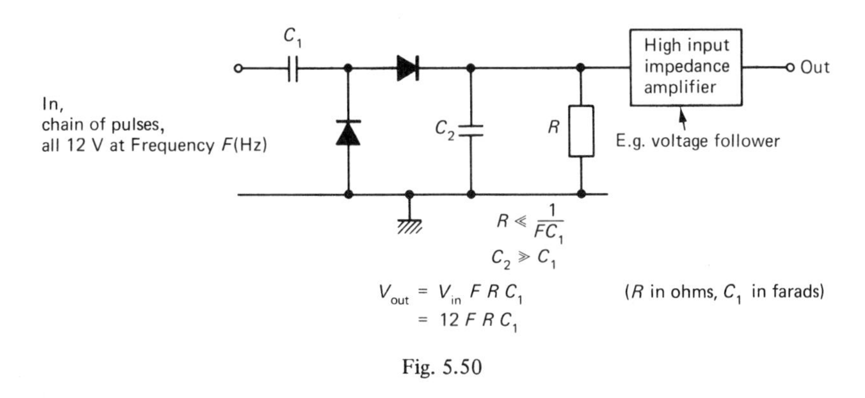


Fig. 5.50

rate of input pulses. The action of this circuit may become clearer if the voltage across the capacitor is thought of as the water level in a leaky bucket, fed by a dripping tap.

## TAPE RECORDERS

Magnetic tape recorders provide a useful means of recording electrical signals. Two-channel recorders ('stereo') are readily available and can be used to record two relevant signals simultaneously, such as the rates of firing of two different nerves, or the changes in a stimulus and in a behavioural response.

The recording system involving moving a magnetisable tape past a fluctuating electromagnet is inherently unresponsive to slow changes in signal (nothing less than 60 Hz gets recorded) and to very fast ones (most tape recorders are unreliable above 10 kHz). The lack of response to low frequencies is one of the tape recorder's main drawbacks, the other being that tapes of biological events all sound much alike and extremely careful labelling and cataloguing is essential if muddles are to be avoided. Various tricks are available to make tape recorders respond to DC or low frequency signals.

### Amplitude modulation

First you pick a frequency to which the tape recorder responds well, say 5 kHz, then you make up a code. Table 5.2 is an example of a code suitable for input signals which vary slowly between +5 and −5 volts.

Table 5.2

| Input signal (volts) | Coded version (volts) | |
|---|---|---|
| −5 | 0.1 | peak to peak amplitude 5 kHz |
| −4 | 0.2 | " |
| −2 | 0.4 | " |
| 0 | 0.6 | " |
| +5 | 1.1 | " |

The tape recorder gain controls are adjusted so that the replay level is the same as the input level. Hence if you input a 0.2 volts/peak to peak signal at 5 kHz while recording, you will get the same voltage back from the tape recorder output when you replay.

An electronic circuit, called the modulator, is used to encode the input signal for recording and another, the demodulator, to decode it again on replay (figure 5.51). The first circuit in the modulator is an adder, which inverts the signal and adds 6 volts to it to make sure that all permitted input signal voltages end up as negative voltages. Next comes the modulator amplifier whose input is shorted to ground each time the output of a 5 kHz oscillator (see chapter 8 for circuit) is in the − 12 volt state. This amplifier reinverts the signal and converts it to a series of brief pulses of appropriately varying height. Lastly, a voltage divider reduces the modulator output to a tenth to suit the tape recorder input circuits.

The pulses are recorded on tape, and, because of the lack of a DC response, reappear on replay as shown in graph (d) equally balanced on either side of the zero-volt line. The demodulator starts with a ×20 gain inverting amplifier. The gain requirement is made up of a ×10 amplifier to compensate for the voltage divider at the end of the modulator and a ×2 amplifier because the next stage is going to lose the half of the signal above the zero line. As the actual information is being carried by the line joining the tips of the pulses, and this is symmetrical above and below the zero line, inverting the signal at this stage has no effect. Next a diode removes the half of the signal above the line, and capacitors smooth out the remainder, just as in a power supply. Finally an adder puts back the 6 volt shift you first thought of, and inverts the signal yet again to make it the right way up.

Theoretically it is possible to use this modulator system for signals for DC to

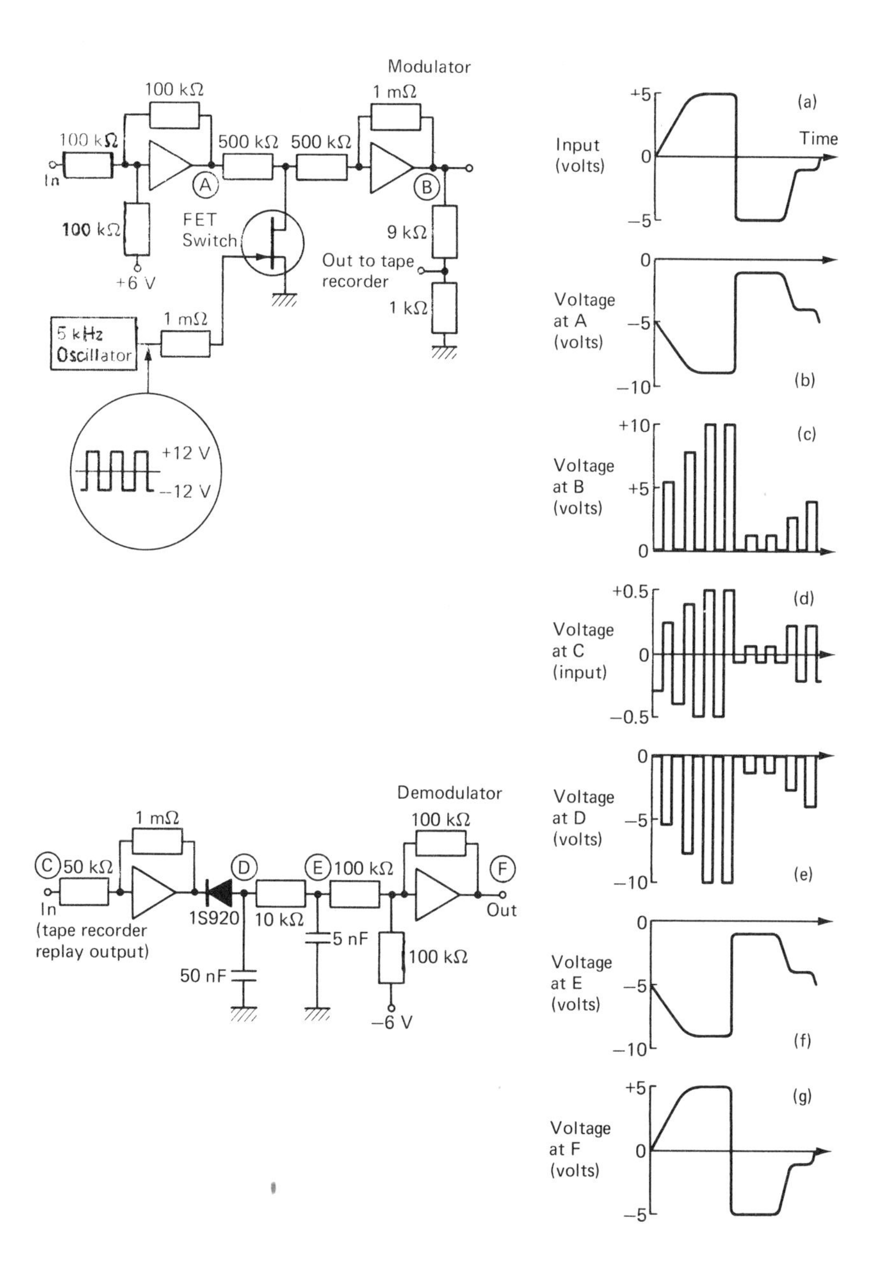
Modulator
100 kΩ
1 mΩ
100 kΩ
500 kΩ
500 kΩ
In
A
B
100 kΩ
FET
Switch
9 kΩ
+6 V
Out to tape
recorder
1 kΩ
1 mΩ
5 kHz
Oscillator
+12 V
−12 V
Demodulator
1 mΩ
100 kΩ
C
50 kΩ
D
E
100 kΩ
F
In
(tape recorder
replay output)
1S920
10 kΩ
Out
5 nF
100 kΩ
50 nF
−6 V
Input
(volts)
+5
0
−5
(a)
Time
Voltage
at A
(volts)
−10
(b)
Voltage
at B
(volts)
+10
(c)
Voltage
at C
(input)
+0.5
−0.5
(d)
Voltage
at D
(volts)
(e)
Voltage
at E
(volts)
(f)
Voltage
at F
(volts)
(g)

Fig. 5.51

2.5 kHz (you need 2 cycles of the 5 kHz 'carrier' to be sure something is there). This bandwidth would mean that a modulated signal fed into the tape recorder would contain sine-wave components ranging from 2.5 kHz to 7.5 kHz and so the smoother in the demodulator would have to act very suddenly after 2.5 kHz if there were not to be lots of 'ripple' from the modulator mixed with the output signal. This is quite difficult to arrange satisfactorily, and most people would settle for a bandwidth of DC to 500 Hz. This code is called amplitude modulation and is used in medium-wave radio transmitters and receivers.

Its disadvantages for our purposes are that tape recorders often have rather a low dynamic range; this means the number of amplitude steps you can reliably separate is small, and hence the output is often rather a distorted version of the input. Moreover slight changes in the speed of the tape and in the gains of the tape recorder internal amplifiers will have serious consequences. But if for instance the signal to be recorded is a marker to indicate whether a stimulus is present or absent, this system is ideal.

## Frequency Modulation (FM)

A code needing slightly more complicated electronics is called frequency modulation (table 5.3). In this system the amplitude of the recorded signal is irrelevant, so it is usual to use the maximum permissible signal levels. The circuit used as an encoder

Table 5.3

| Input (volts) | Coded version (kHz) |
|---|---|
| −5 | 2.5 |
| −4 | 3 |
| −2 | 4 |
| 0 | 5 |
| +5 | 7.5 |

is a close relative of the timebase circuit described in chapter 4. It consists of an operational amplifier integrator whose capacitor is discharged each time the output reaches − 6 volts. As the input voltage varies (+ 5 volts is added to all inputs to set the frequency range properly) the current flowing into the integrator capacitor varies and hence so does the frequency with which it needs resetting (figure 5.52). The sawtooth-shaped output voltage curve can be used directly as a tape recorder input, but purists first turn it into a square wave by AC-coupling into a second comparator.

The demodulation of the played-back signal is straightforward. A comparator gets us back by the square-wave signal. The positive-going pulses output by the comparator are fed into a rate meter circuit (figure 5.53). The final stage of demodulation involves subtraction of the 5 volts initially added to the signal in the modulator circuit. As this is going to involve an adder which will invert the signal

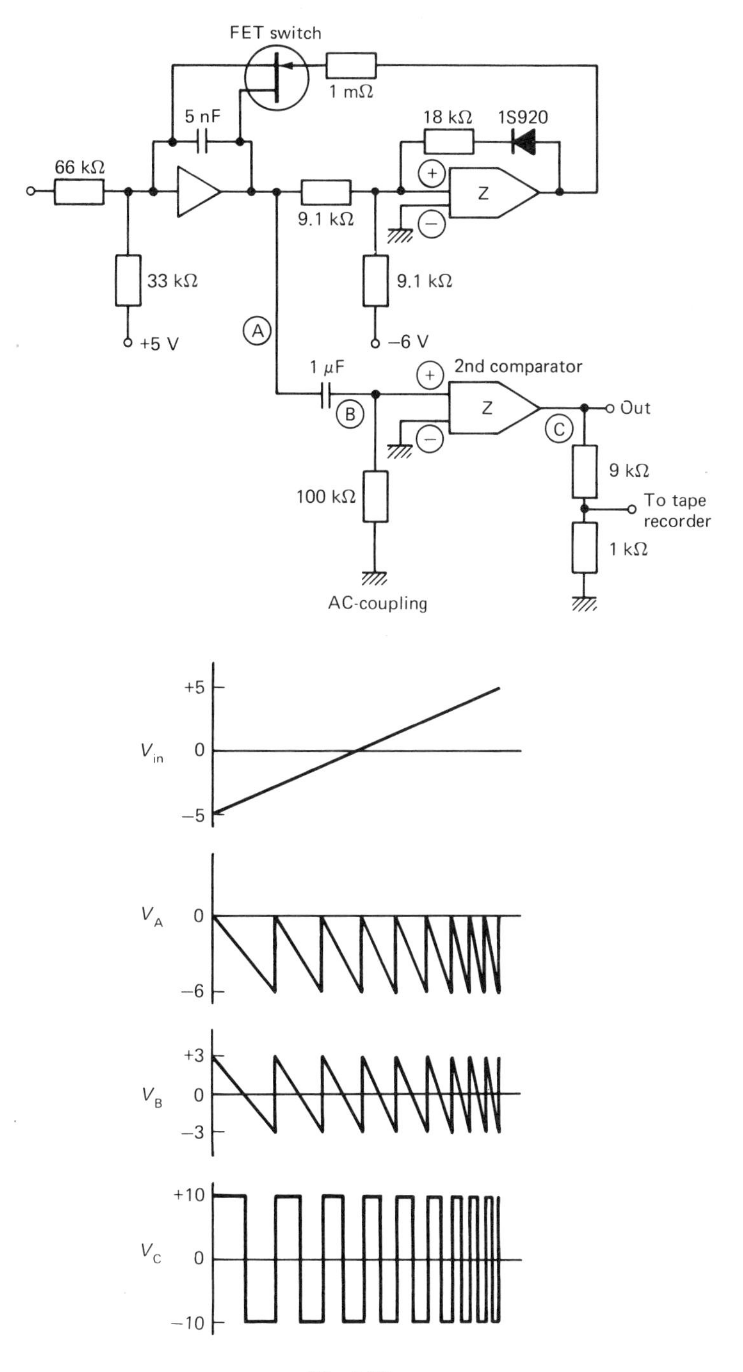

FET switch
1 mΩ
5 nF
18 kΩ
1S920
66 kΩ
Z
9.1 kΩ
33 kΩ
9.1 kΩ
+5 V
A
−6 V
1 μF
2nd comparator
B
Z
Out
C
9 kΩ
100 kΩ
To tape recorder
1 kΩ
AC-coupling
+5
$V_{in}$
0
−5
$V_A$
0
−6
+3
$V_B$
0
−3
+10
$V_C$
0
−10


Fig. 5.52

it is necessary to arrange a downwards staircase (by reversing diodes) so that our ultimate output signal finishes the right way up.

The frequency modulation coding system (used in VHF transmitters) is totally unaffected by variations in the gain of the tape recorder amplifiers. FM tape recorders with bandwidths from DC to 2.5 kHz and four separate recording channels

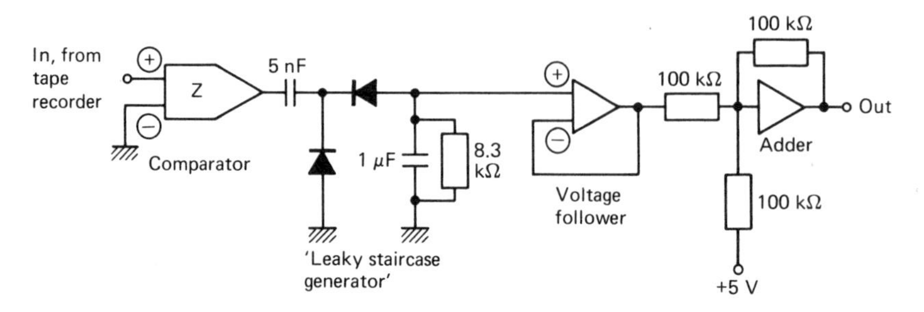


Fig. 5.53

are commercially available. The snags with this system are that it is very much affected by changes in the speed of the tape, and that if the demodulator is given no input (this happens if there is a small gap in the surface of the recording tape called a dropout) its output goes down at once to − 5 volts. So dropouts produce artefacts looking just like nerve impulses!

## Mark–space Coding

The last and most sophisticated coding system is mark–space ratio coding (figure 5.54). The time intervals between the upstrokes of this square-wave signal are all equal. The downstrokes, however, can vary in their position, which is specified by the ratio

$$\frac{\text{'mark' time interval}}{\text{'space' time interval}}$$

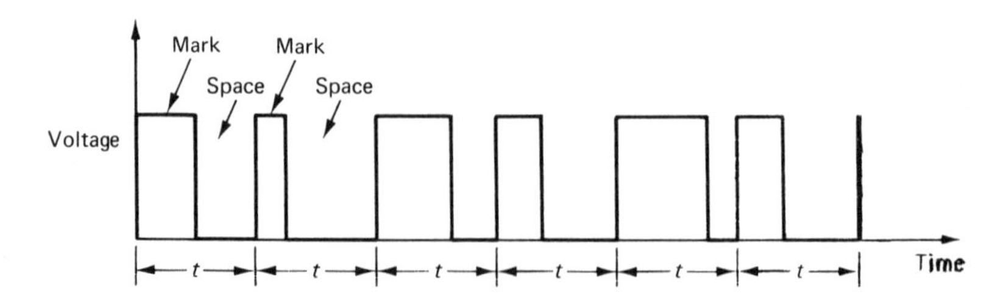


Fig. 5.54

The position of the downstroke is made to depend on the input voltage (table 5.4). Provided that tape recorder speed variations are slow compared with the frequency of the coded pulses they do not affect this system; the mark–space ratio which carries the information stays the same even if the tape speed is halved. The

Table 5.4

| Input (volts) | Coded version |
|---|---|
| –5 | 1:10 |
| 0 | 1:1 |
| +5 | 10:1 |

decoder and encoder circuits for this system are too complex for this book. The disadvantage of this system is that it has a very limited bandwidth, generally DC to 100 Hz with domestic tape recorder systems.

## 5.5 COMPUTER DATALOGGERS

The data recording systems dealt with so far in this chapter rely on tape recorders, electronic event-recognisers, and counters. This approach works extremely well for the classical biological experiment in which you want to record only the imposed change in a single environmental variable, say light intensity, and a single set of response events, say spikes all coming from one nerve fibre. One can use the two channels of a domestic stereo tape recorder (cost around £70). The electronics required are straightforward, probably an FM coder for one channel to record steady light levels, and a window discriminator to sort out the relevant spikes. If, however, you want to record the responses of, say, two nerves and a muscle to changing light intensity, you need a four channel instrumentation recorder (cost around £500) and a great deal more electronics. For eight separate recording channels things become unreasonable. Tape recorders to cope with this use half or one inch tape and cost several thousand pounds. Really complex sets of the building block circuits described earlier are very difficult to get to work properly because of complicated interactions between circuits.

This situation most frequently occurs when dealing with human subjects, when it is essential to extract as much information as possible in the limited time your subjects allow you to work on them. It is not uncommon to monitor heart rate, respiration, skin resistance, eye position and several electroencephalograph channels simultaneously. The traditional method of dealing with all this data is the multichannel pen recorder, which leads to an enormously laborious data analysis with ruler and pencil. The alternative is to use a computer datalogger. These centre around a digital voltmeter, which displays the voltage across its input terminals as a set of illuminated digits using electronic number-tubes. This works by counting pulses from an oscillator till the count reached corresponds to the input voltage, when the counting is stopped and the answer is displayed. This correspondence is determined by a comparator which looks at the difference between the input voltage, held steady by a sample-and-hold circuit while the count proceeds, and a voltage directly proportional to the number the count has reached. The latter is generated by a device called a digital to analogue (D–A) convertor (figure 5.55).

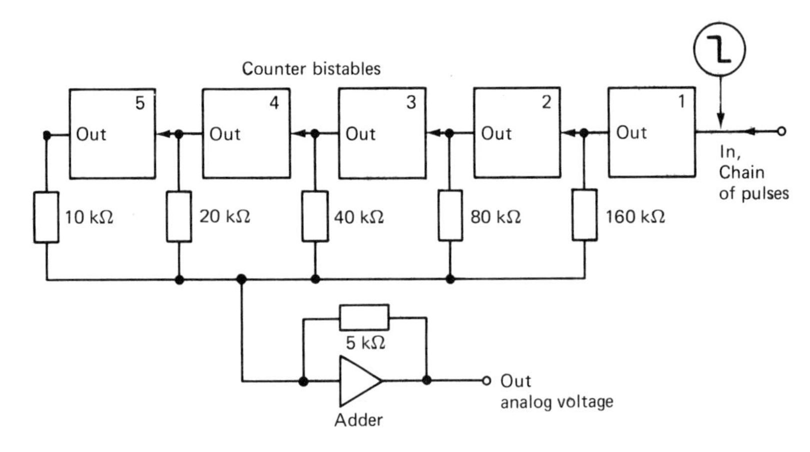


Fig. 5.55

This is an adder with input resistors connected to the outputs of each of the counter bistables. If all the bistable outputs are at zero volts, the adder output will be at zero volts. If a single pulse is input to the counter, the output of bistable one goes to + 10 volts, and the adder output to $-10 \times 5/160 = -0.313$ volts. The next input pulse returns bistable one to zero volts, and the negative-going change at the output of bistable one makes bistable two's output go to + 10 volts. The operational amplifier's output is now $-10 \times 5/80 = -0.625$ volts. The next input pulse puts bistable one back to + 10 volts, so the adder receives inputs down both the 80 kΩ and 160 kΩ resistors. Its output is hence − 0.938 volts. Each subsequent pulse adds − 0.313 volts to the adder output voltage, until the counter reaches binary 11111 (i.e. 31) when the output voltage reaches 9.688 volts. The next impulse returns all the bistables, and the adder output, to zero volts.

The computer datalogger uses the digital voltmeter to convert the input voltage from the experimental setup to a binary number. It then records this binary number as a pattern of holes and spaces on a paper tape, which can later be fed into a digital computer, which in turn can be programmed to look for specified events and correlate them on many channels. Paper-tape punches come in three sorts, slow (10 sets of eight digit binary characters each second) medium (110 characters per second) and fast (240 characters per second). Fast punches are very expensive, so most people use the medium-speed ones. These are incredibly noisy machines and really need a separate room to themselves.

Many separate input channels are dealt with by sampling the first channel (say heart rate), punching a number, sampling the second channel (breathing rate), punching a number, sampling the third, and so on in rotation. This process of sharing a single-channel output (the series of numbers on the paper tape) among many input channels is called multiplexing. The computer later sorts out the interleaved sets of data into separate lists for each input channel. If the multiplexer integrated circuit ever makes a mistake and skips over a channel all the rest of the data will be assigned to the wrong channel, so it is advisable to insert some distinctive marker at the beginning of each series of readings. If it can be guaranteed that none of the input channels will ever read zero, then an extra channel reading zero volts can be inserted. If this is not the case it will be necessary to sacrifice one of the binary digits. Eight hole tape enables us to record numbers up to binary 11111111 (255). If we want to use binary 10000000 as a marker we must restrict the numbers being recorded to ones less than binary 1111111 (127). Hence if we have ten input channels, each channel will be sampled eleven times a second, and the voltage found there recorded with an accuracy of about 1 per cent. This is clearly only going to be an adequate record of slowly changing inputs, for the effective bandwidth is only DC to 1 Hz. It is necessarily the case with multiplexing that the more channels you have sharing a given output channel the less the bandwidth which can be alloted to each.

This kind of datalogger is enormously useful for recording slowly changing voltages on many channels. If the voltages are changing very slowly, it may save a lot of paper tape and computer time to punch the time it takes the voltage to change by, say, 1 per cent, or the time between electronically recognisable events, and accompany each punched reading with a punched channel-number, rather than punching the voltages on each channel in rotation at full speed. Having the results emerge from a computer in the form of a neatly printed table of statistical significances clearly has its attractions, but the programs required for this do take a long time to write and to get properly working, and I think the effort is almost never justified for a single experiment. If you are going to use the same set up for very many experiments (e.g. in a complex bioassay technique), then it is very worthwhile indeed to spend the time and effort needed to get a completely automated data collection and analysis system working.

This account of computer datalogging assumes the usual sort of budgeting available for biological research. If money is no object and it is possible to use a computer with direct electrical connections to the experimental set-up, then very high rates of data transmission are possible.

# 6 Controlling Stimuli and Measuring Responses

## 6.1 LIGHT

There are two main sorts of lamp, the fluorescent strip lamp and the tungsten filament lamp.

### Fluorescent Strip Lamps

A strip light has three components, the tube, the choke (a big coil with an iron core) and the starter (figure 6.1). When the mains is connected to the system, the neon in the starter glows and heats up the bimetal strip which bends till the switch closes

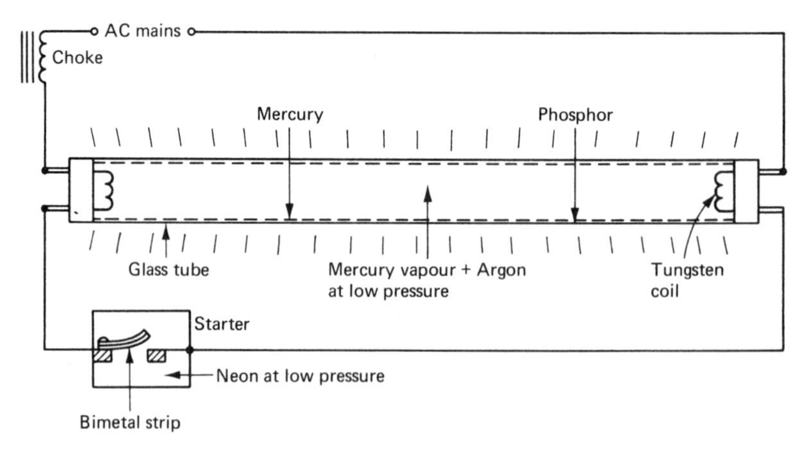


Fig. 6.1

(see figure 6.26). Current now flows through the two tungsten coils in the lamp tube, which rapidly become red hot. By this stage the bimetal strip in the starter has cooled, and the contact is broken.

The inductor (choke) in the circuit resists the change in current flow and supplies a high voltage pulse between the two hot tungsten coils causing the lamp to 'start' – electrons are attracted along the length of the tube towards the tungsten coil at the more positive voltage, ionising the argon, which keeps things going, and making the mercury vapour emit ultraviolet light, which the phosphor on the walls of the tube turns into visible light. A column of ionised gas tries to drag in as much current as possible, but the inductor prevents things going too far.

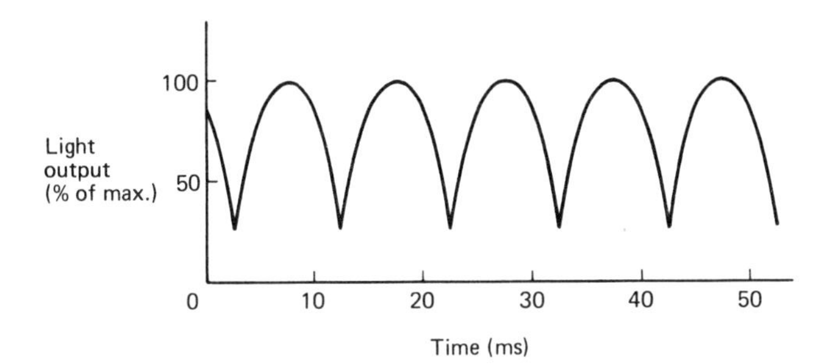


Fig. 6.2

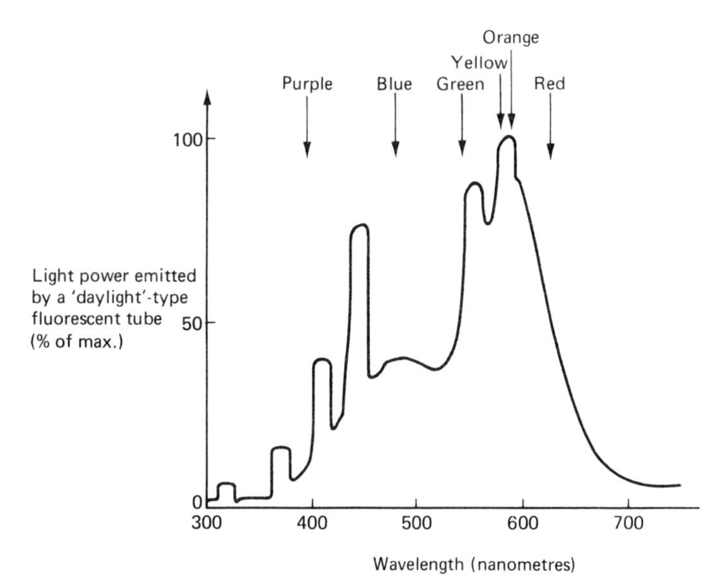


Fig. 6.3

Fluorescent tubes flicker at 100 Hz, twice for each cycle of the AC mains (figure 6.2). The colour of the light depends both on the phosphor and the visible spectral lines of mercury (figure 6.3). The unit of light colour is the wavelength of the light concerned in nanometres (metres $\times 10^{-9}$). The great advantage of fluorescent tubes is that you get a lot of fairly white light (and not too much heat) for a low electrical power consumption and over a long life. The disadvantages are

(1) The flicker may possibly affect some biological systems, and cannot easily be removed by running the tube on a DC supply. (It does work provided you connect a resistor in series with the choke, but all the mercury congregates at one end and the tube has to be turned round every eight hours.)

(2) The light output declines as the tube ages (figure 6.4).

(3) The light output depends on the temperatures of the glass of the tube (figure 6.5).

(4) Physically big light sources are unsuitable for supplying light to optical system (e.g. projectors) which need something approximating to a point source.

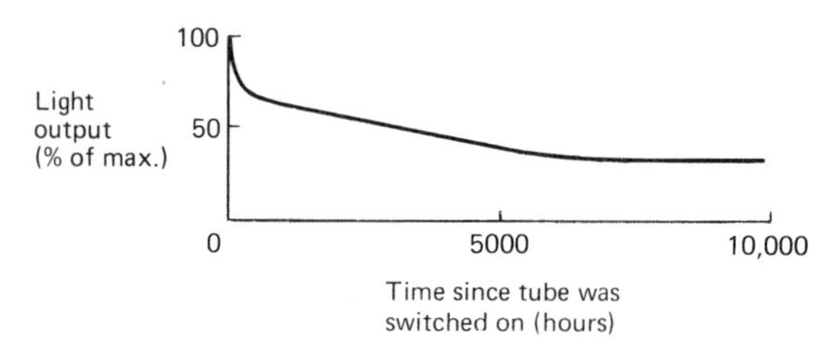


Fig. 6.4

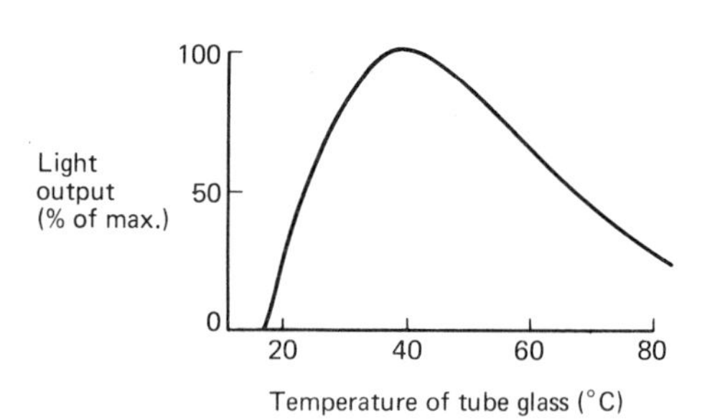


Fig. 6.5

The principal application of fluorescent tubes is in the production of very large bright stimulus zones, suitable for mimicking the sky on a bright day (e.g. for experiments involving animal phototaxes, or for growing plants). To approximate daylight the stimulus area needs to be nearly solid with fluorescent tubes. If high light intensities are needed in a temperature-controlled area it is a good idea to keep the chokes and starters outside the constant temperature room.

## Tungsten Filament Lamps

Tungsten filament lamps are the most commonly used light sources. They provide a completely steady light provided they are run from a DC supply. (The AC mains does produce some flicker – figure 6.6.)

If the voltage across a tungsten filament lamp is increased by 5 per cent from the rated voltage, the filament gets hotter, the light emitted becomes bluer, brighter (by 20 per cent) and the lamp life is reduced (by 50 per cent). If the voltage is decreased by 5 per cent the light becomes redder, its intensity is reduced by 15 per cent, and the life of the lamp nearly doubled. In the design of the 240 volt tungsten filament

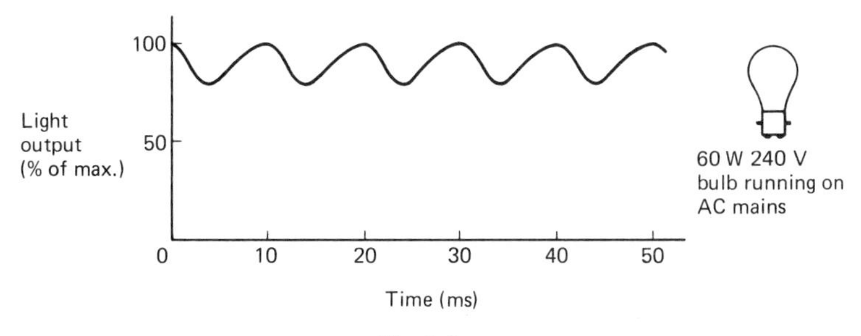


Fig. 6.6

lamps commonly used for domestic lighting, long life is of a paramount importance, so filaments are kept comparatively cool, and lives of 1000 hours are achieved. Projector lamps and microscope lamps need to be bright and white, so the filament must be hotter. Their lives are thus much reduced: 240 volt projector lamps last only 70 hours, and their light output falls drastically (by up to 40 per cent) towards the end of their life because tungsten from the filament is evaporated onto the glass of the bulb.

Some of the undesirable features of projector lamps can be removed by putting a little iodine into the lamp. The iodine vapour keeps the inside of the bulb (which is small, made of quartz, and very hot) clean, and redeposits the evaporated tungsten back onto the filament. The light is much less orange and lamp life is extended to

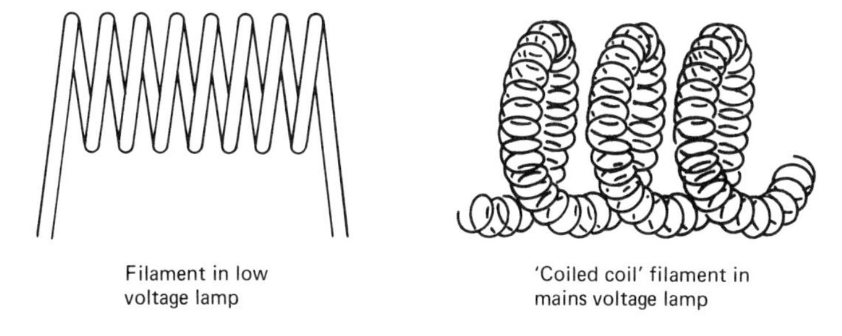


Fig. 6.7

150 hours. There is very little fall off in light intensity as the bulb ages. These tungsten–halogen projector bulbs are generally for 12 or 24 volt operation, so smooth stable DC power supplies are easy to arrange, and in many ways they are ideal for use in experiments. Their only disadvantage is that the light they emit is quite

strongly polarised. This is a property of all lamps in which the filament consists of straight parallel wires (i.e. most low voltage lamps) and does not occur with the coiled or doubly coiled filaments used in 240 volt lamps (figure 6.7).

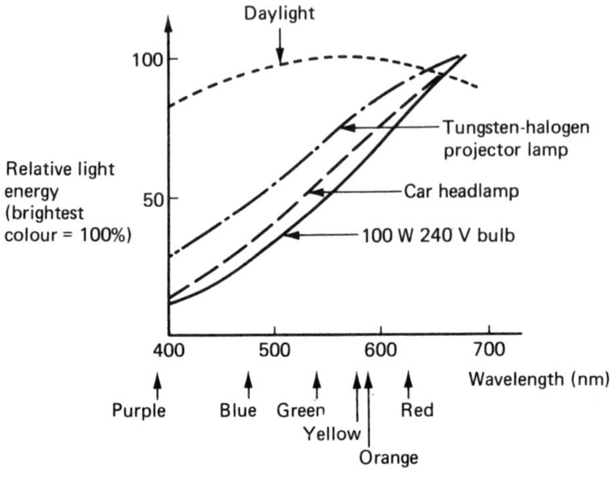


Fig. 6.8

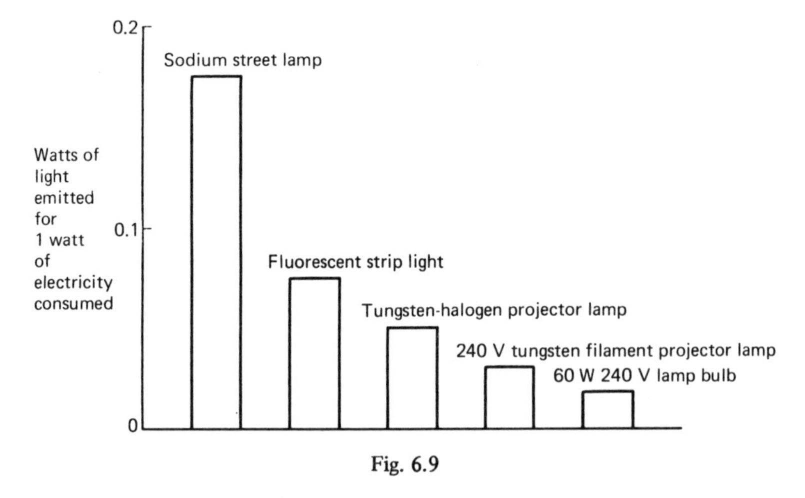


Fig. 6.9

The distribution of the light energy between the colours for various sorts of lamp is given in figure 6.8.

In fact in all lamps the majority of the electrical energy is converted to infra-red light and heat (figure 6.9).

## Presenting Visual Stimuli

It is often important to control the shape, colour, duration and movement of patterns seen by an experimental animal. The arrangement I use for presenting stimuli is based on a small slide projector, which has an unusually accessible space behind the lens in which filters can be put (figure 6.10). The colour wedge is a long narrow filter which looks like a slice of rainbow when held up to the light. Its colours are pure because they are made by interference (the optical process which colours oil slicks on puddles), rather than by pigments. The slit shines light through

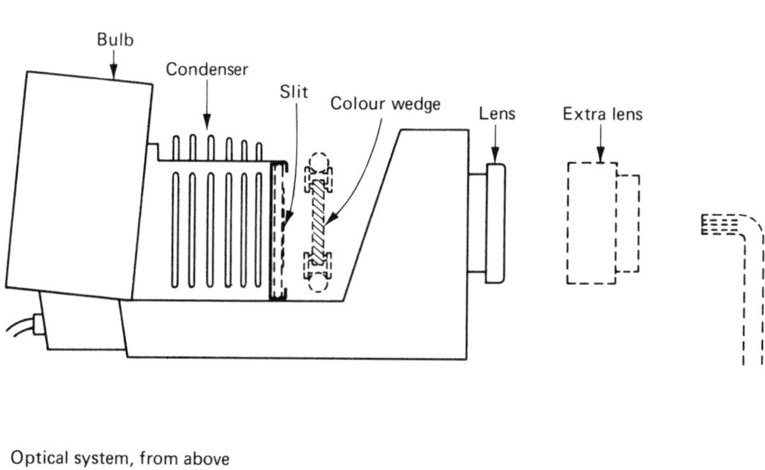


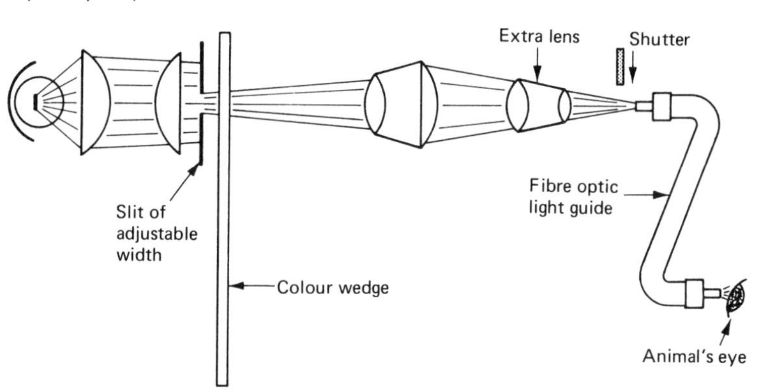


Fig. 6.10

a small portion of the wedge, and the colour is altered by sliding the wedge past the slit.

The projector lens and the extra lens focus the coloured light onto the end of a light guide, a flexible tube which carries the light to the experimental set-up and removes the need for precisely aligning the whole projector. Light guides are bundles of light-conducting fibres packed tightly down the centre of a black plastic tube. Each individual fibre has a core of high refractive index glass or plastic and a sheath of lower refractive index, and guides light round corners by repeated total internal reflections (figure 6.11). The need for total internal reflection means that the cone of light at the input end of the light guide must be fairly shallow, of angle not more than 50°. A 50° cone of light will emerge from the output end (light

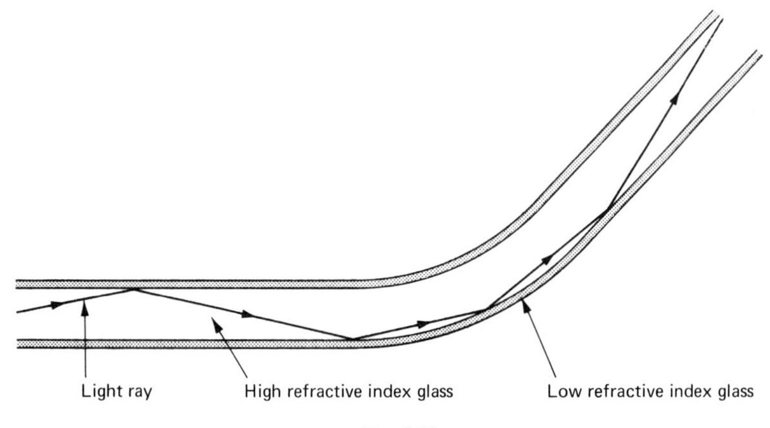


Fig. 6.11

guides are completely reversible, either end can be used as input or output). Various sorts of fibre-optic light guides are available.

(1) Plastic, which is extremely cheap and can be cut to any required length with scissors. The coarse fibres (0.25 mm diameter) are easy to arrange in any suitable shape (e.g. a straight line) at the output end.

(2) Glass, the fibres of which are much finer (0.05 mm diameter) and with the ends of the bundles Araldited into fixed configurations (generally circles, but slits are available) by the manufacturer.

(3) Quartz, which look like glass-fibre light guides, but transmit ultraviolet light.

The light-transmitting abilities of each type of light guide is shown in figure 6.12.

The bundles described so far have randomly arranged fibres, so that they will transmit a patch of light but not a picture. It is possible to buy glass-fibre optics (at around £700 for a six-foot length) in which the ordering of the fibres is preserved; an image formed at one end of the system will clearly be visible when viewed from the other end (figure 6.13).

The duration of the light stimulus is controlled by a shutter which interrupts the light beam, preferably at the input end of the light guide. Camera shutter mechanisms should be avoided, for they are very variable in their timing, and difficult to set off with an electrical signal. I have had considerable success with cardboard vane stuck to the moving bit of a relay; the latter is an electromagnetic switch in which a current in a coil turns its iron core into a magnet, which in turn attracts the iron toggle of a sensitive switch (figure 6.14). With the small relay shown this shutter works well for times down to 20 milliseconds. Suitable electrical pulses to drive the relay coil are obtained from a delay circuit and a transistor switch (see figure 5.42). The relay switch contacts play no part in its functioning as a shutter, but can be used to indicate exactly when the shutter is open.

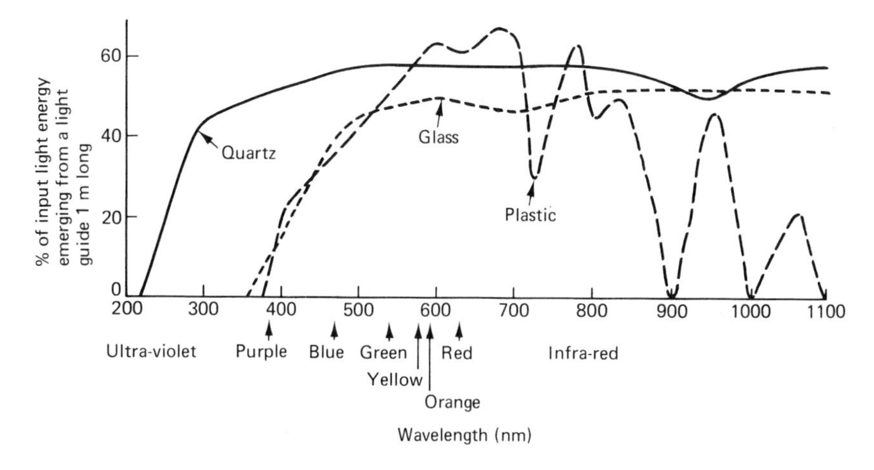


Fig. 6.12

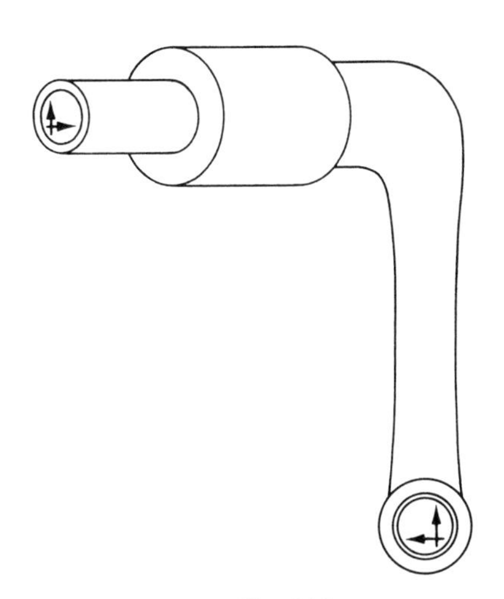

Fig. 6.13

For really brief flashes of light the simplest technique is to use an oscilloscope tube as a light source! Switch off the timebase, so the bright spot stays on the vertical centre line of the tube. Apply a voltage pulse to the $Y$-amplifier of height just sufficient to move the spot vertically to a position beneath the end of a light guide pressed against the tube face. The pulse length controls the time for which the other end of the guide lights up, which can be less than one microsecond if required. The

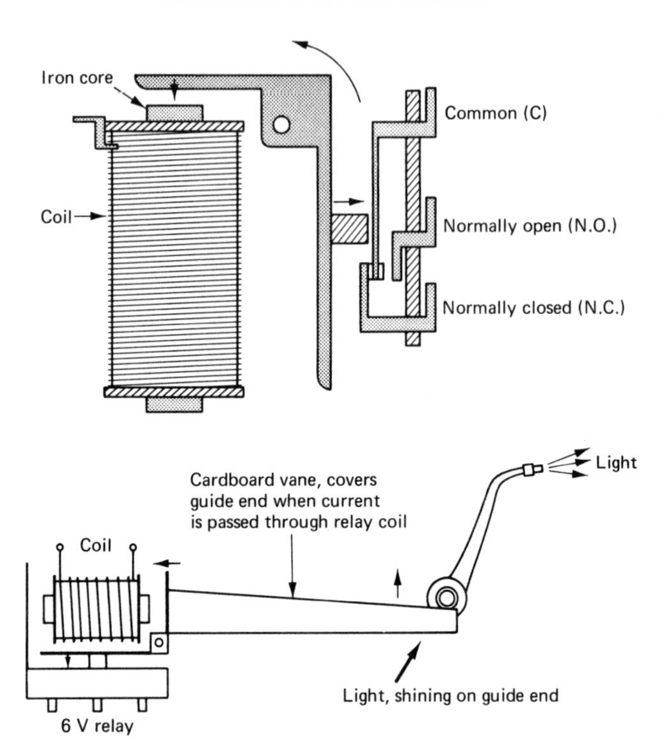


Fig. 6.14

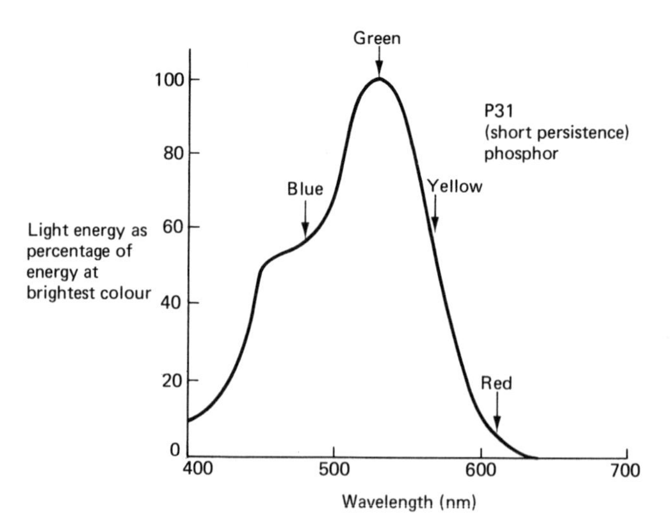


Fig. 6.15

oscilloscope spot looks bright green, but in fact contains a reasonable amount of light of most colours (figure 6.15). Coloured stimuli can be produced by interposing a filter between the light spot and the guide end.

## Light Measurement

(*i*) *Absolute units of light measurement.* The only reliable way to assess the energy of a light stimulus is to set up the precise conditions of the experiment with a suitable photocell substituted for the eye of the experimental animal. The units of light intensity are watts per square metre. The sensible thing to quote in most biological contexts is the illumination of the experimental animal's eye (figure 6.16).

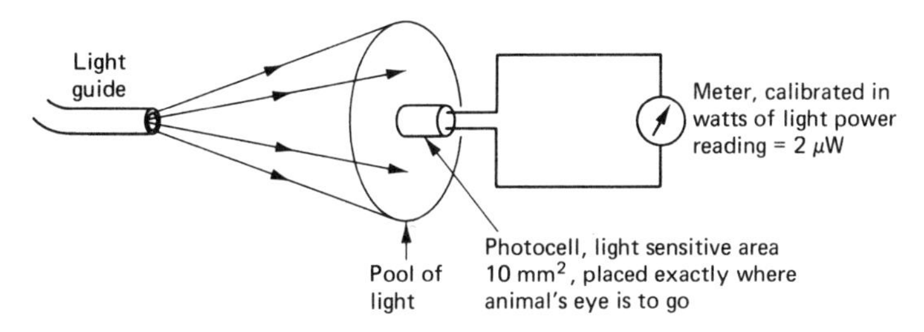


Fig. 6.16

The meter registers that the illumination is $2\,\mu$ watts/10 mm$^2$, i.e. $2 \times 10^6/10 =$ $2 \times 10^5$ $\mu$watts/m$^2$, = 0.2 watts/m$^2$. All measurements of light intensity must be accompanied by information about the light colour. The unit of light colour is the wavelength of the light concerned and is usually expressed in nanometres (metres $\times$ $10^{-9}$). With pure colours produced by interference filters it is sufficient to state the colour bandwidth the manufacturer says is transmitted by the filter; for example if the light in the example above had passed through a blue interference filter, its full description might read: 0.2 watts/m$^2$, 450–460 nm.

If the light is white, or is a combination of many colours, it is best to plot a graph of the colour content (for example, figure 6.17). This is done either by the use of a

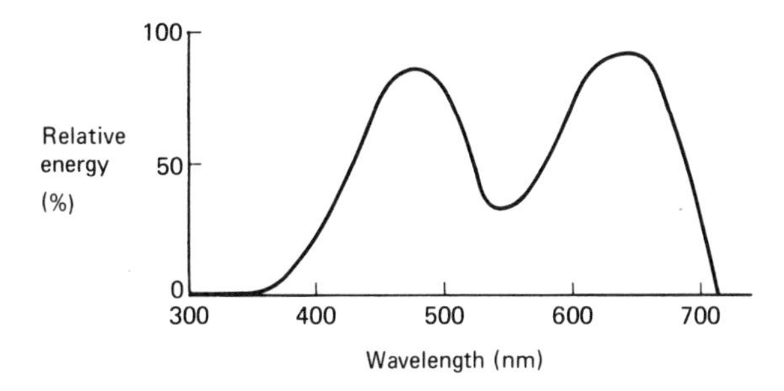


Fig. 6.17

special machine called a spectrophotometer, or by taking readings with the photocell pointing through each of a series of interference filters, going right through the colour spectrum. Each reading must first be multiplied by 100/(percentage transmission) of the filter through which it was taken.

Things to avoid in photometry are:

(1) Units of luminance, which express the brightness of a light source. They are fraught with solid angles and imaginary perfectly diffusing surfaces.

(2) The light engineer's units (e.g. lux, lumens, foot lamberts, apostilbs) which are obtained by multiplying the light energy in watts for each colour present by the sensitivity of the human eye to that colour, then adding up the results and multiplying by various constants. These are entirely appropriate to calculating good light levels for offices, but totally inappropriate for experimental work on animals whose colour sensitivity differs from that of the human observers.

(*ii*) *Choosing a photocell.* The above account of measuring a light intensity assumes the photocell used responded equally to light of all colours. The only sort of photocell which fits this requirement is the vacuum thermopile, which is basically a layer of soot on a sensitive temperature-measuring device. The more light the soot absorbs, the hotter the temperature measurer gets, and the greater the current it generates. The output current is directly proportional to light intensity, and the manufacturers tell you how many microwatts of light power absorbed each microamp of output current represents.

The devices are expensive, fragile, unstable and require very sophisticated meters or amplifiers because their output currents are very small. They respond to radiation well beyond the visible at the infra-red end of the spectrum, and in general they get more enthusiastic about a hot soldering iron than a light bulb.

The main use of thermopiles is to calibrate more robust and sensitive photocells, which however do not respond equally to all light colours. The calibration procedure involves using a set of interference filters to produce light of a series of just non-overlapping pure colours going right through the spectrum, and measuring the light intensity for each colour with the thermopile. The cell being calibrated is then substituted for the thermopile (it must go in exactly the same place) and its output current measured for each light colour.

It is now possible, taking account of the ratio of the size of the sensitive area of the photocell to that of the thermopile, to work out a table of how many microamps of photocell output current are produced by one microwatt of light power falling on the cell for each separate colour. The newly calibrated cell is always subsequently used on conjunction with this table. To measure the intensity of a light of pure colour you simply divide the cell output current by the factor from the calibration table appropriate to the colour concerned.

To measure the intensity of a white light it is necessary to take photocell readings with each of the series of filters (going right through the spectrum) placed in front of the cell. The intensity contribution for each colour can be calculated using the appropriate factor from the calibration table, and remembering to multiply the answer by 100/(percentage light transmission of filter). The total light intensity can now be found by adding up all these contributions.

The best sort of photocell to use for everyday photometry is a vacuum photocell such as the Mullard 90AV. Figure 6.18 gives details of its construction and the

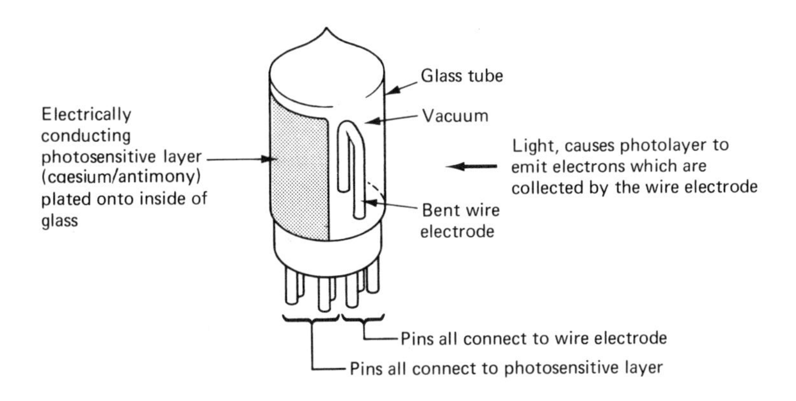


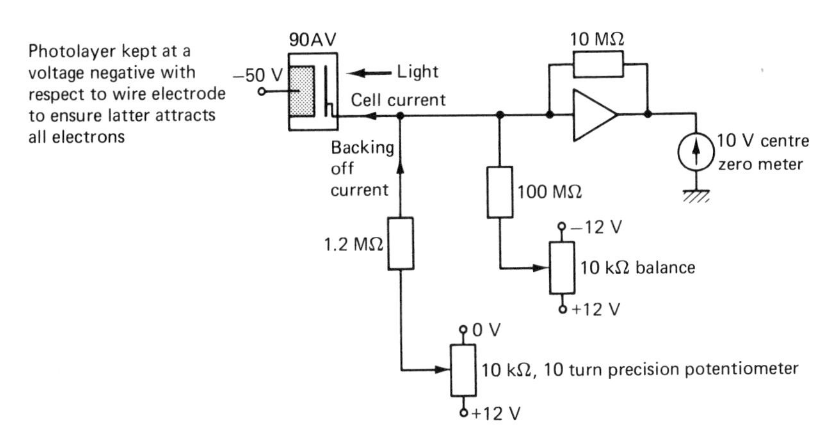


Fig. 6.18

circuit used to measure its current. Figure 6.19 shows its spectral response curve. The peak of the curve can fall anywhere within the range indicated, the exact position depending on the parameters of the manufacturing process.

To measure the photocell current, set the precision potentiometer to its zero volt end, cover the photocell to prevent any light entering and adjust the balance knob to zero the meter. This allows for any 'dark current', which is the tiny signal from the photocell which remains even in complete darkness and is largely caused by the radioactive decay of the glass tube. Then uncover the photocell and turn the precision potentiometer till the meter returns to zero. The backing off current from the potentiometer (see chapter 4) now exactly equals the photocell current, which can hence be read off from the scale attached to the precision potentiometer.

An alternative to the vacuum photocell is a recently produced range of precisely calibrated photodiodes. These are semiconductor devices whose electrical resistance depends on the light intensity falling on them. Figure 6.20 shows the response curve for a silicon photodiode. It has been made linear by filters fitted during manufacture.

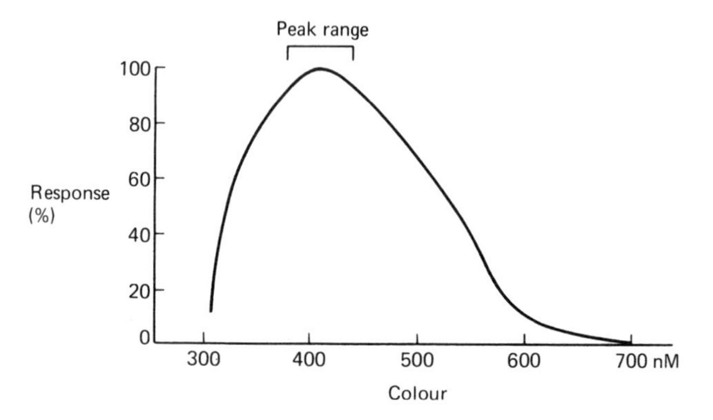


Fig. 6.19

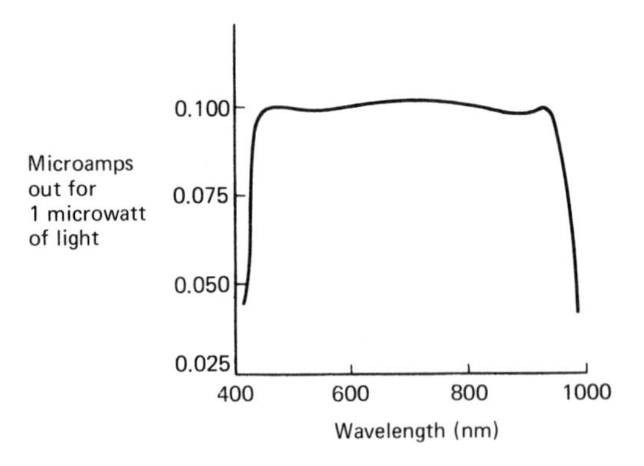


Fig. 6.20

For measuring extremely dim lights it is possible to obtain vacuum photocells with built in high-gain low-noise amplifiers. These are called photomultipliers (figure 6.21). Each time an electron collides with one of the plates (called dynodes) it knocks off several electrons. These are all attracted towards the next plate, which is kept at a more positive voltage by connecting it to a less negative point on the big voltage divider. Hence the process repeats itself, so the final consequence of a single light photon hitting the photosensitive layer and releasing a single electron is a measurable pulse of current at the collector end of the chain of charged plates.

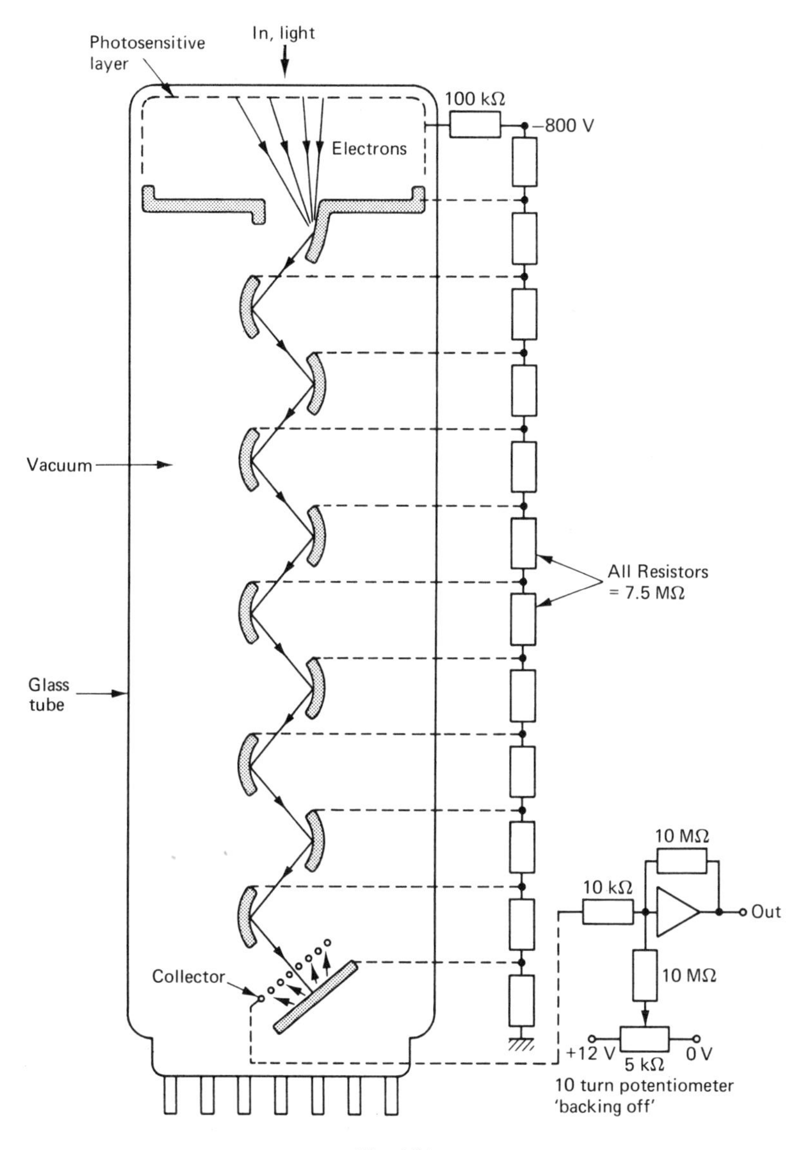


Fig. 6.21

This pulse of current is amplified by an operational amplifier circuit, supplied with a backing off control to compensate for the multiplier's dark current. Photomultiplier tubes are quite expensive (around £20–25), need good magnetic shields made of mumetal (around £10), and very stable high-voltage DC supplies (around £200). Both photomultipliers and vacuum photocells will respond to changing light intensities involving lights flickering at extremely fast rates.

(*iii*) *Other photocells.* There exists a wide range of light-sensitive semiconductor devices falling into two groups: those which actually generate electricity from light, for example the selenium cells used to power space probes; and those which have an electrical resistance which depends on the light intensity to which they are exposed, for example the cadmium sulphide photoresistors and the silicon photodiodes. The latter are often arranged in pairs in the same component so that it does not matter which way round they go in a circuit.

I am dealing with all these components very briefly because I think that in future they will be little used in biological instrumentation. Their photometric usefulness (with the above-mentioned exception of some photodiodes) is limited to determining photographic exposure times, and they are rapidly being superseded by integrated-circuit light detectors (dealt with in the next chapter) for applications involving detecting, recognising and counting objects or animals.

## 6.2 HEAT

Heat measurement does not generally give much trouble because the traditional mercury-in-glass thermometer is such an accurate and reliable instrument. The electronic techniques are useful only in cases where the temperature of a minute area needs to be measured, or when it is convenient to have temperature translated into an electrical signal to work a pen recorder. There are two commonly used temperature transducers, the thermocouple and the thermistor.

### The Thermocouple

Thermocouples are generally made whenever required from a special sort of twincore polythene-insulated wire. The two wires are of different metals; generally one is copper and the other constantan (a 60 per cent copper, 40 per cent nickel alloy) (figure 6.22). This works as a temperature transducer because chemically more active metals hang on less tightly to their electrons, and hence have a greater concentration of electrons at their surfaces, than do chemically less active ones. This means that there is a tendency for electrons to flow across a joint between two different metals, and hence for an electric current to flow. In normal circumstances this is hard to show, because you have to join up the two metal wires somewhere else to complete the electrical circuit, and this second join contributes a counterbalance (figure 6.23). But the extent of the electron's freedom in the two metals is differentially affected by temperature, so if you keep one of the joints hot and the other cold a current does flow round the loop. We have broken the loop to insert a differential amplifier to attain signal voltages suitable for a meter, oscilloscope or pen recorder.

The advantages of thermocouples are that they are simple, cheap, reliable, accurate, give very repeatable readings and work over a huge temperature range (the copper–constantan ones will work from − 100 to + 400 °C, and platinum–rhodium ones work

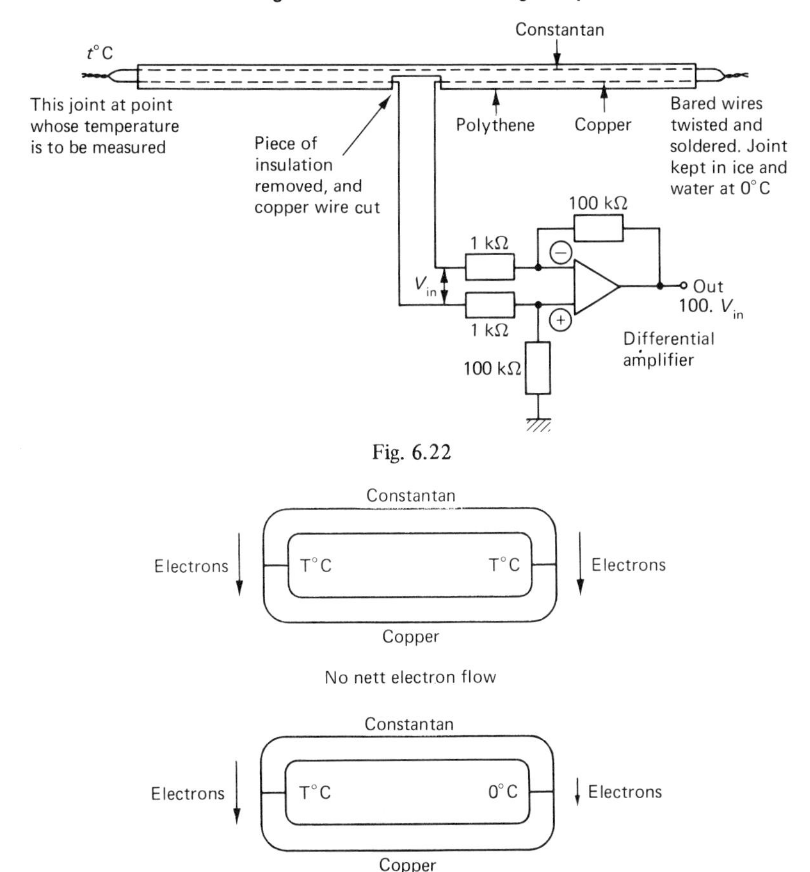


Fig. 6.22



Fig. 6.23

up to 1800 °C). They can be made extremely small and they respond very quickly to temperature changes. However they do need a beaker of ice and water and quite a sensitive amplifier, particularly if you are looking for temperature changes of the order of 0.01 °C, which cause output voltage changes of only 4 microvolts. Table 6.1 gives an indication of their output voltage for different temperature differences.

Table 6.1

| Temperature, (°C) | Voltage, (mV) |
|---|---|
| −10 | −0.38 |
| 0 | 0 |
| 10 | +0.39 |
| 20 | +0.78 |
| 30 | +1.19 |
| 40 | +1.60 |
| 50 | +1.80 |

## Thermistors

Thermistors are pieces of semiconductor materials (e.g. oxides of titanium, manganese, cobalt, or nickel) whose electrical resistance drops dramatically as their temperature is increased (figure 6.24).

The main problem with thermistors is that to measure their resistance you have to pass a current through them, and this heats them up and alters the resistance you

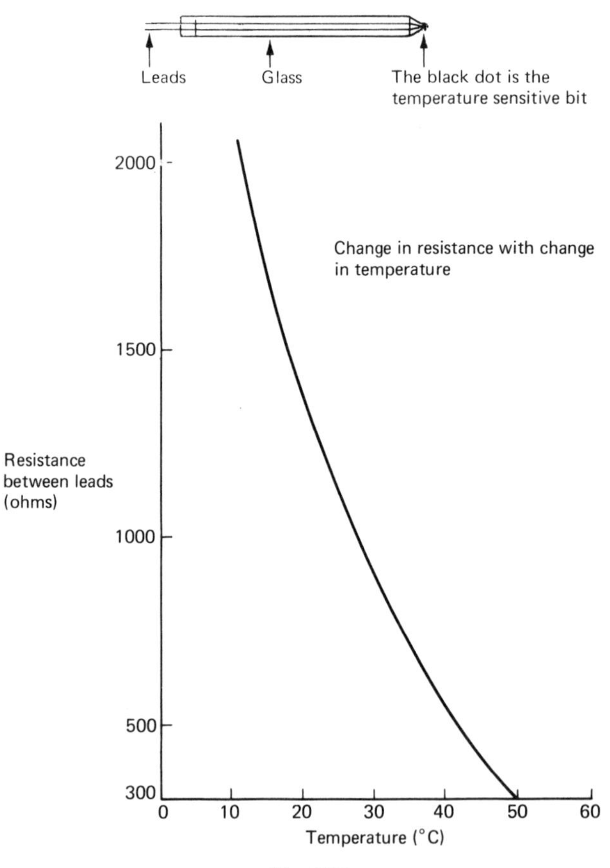


Fig. 6.24

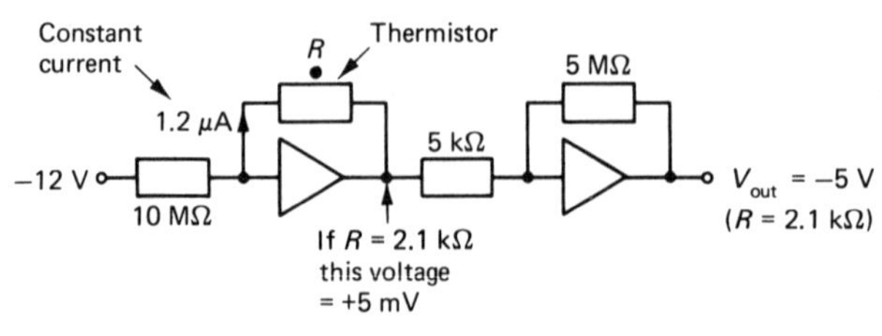


Fig. 6.25

are trying to measure. The measuring current should be kept small and constant – a good circuit is given in figure 6.25. The first amplifier has the thermistor as a feedback resistor and passes a constant very low current through it. The second amplifier has a big voltage gain to produce a useful signal level.

For work in which the thermistor is immersed in large volumes of liquid and thus cannot be heated up appreciably by the measuring current, this electronic sophistication is unnecessary and it is satisfactory to use an avometer to measure its resistance, and hence the temperature of the liquid.

The only disadvantage of the thermistor is that its readings are not so repeatable as those taken with a thermocouple, and the device does need fairly frequent zeroing (e.g. checking that an ice/water mixture gives a 0 °C reading) if accurate measurements are needed.

## Controlling Temperature

The thermostat is the classical device for controlling temperature. It is a switch which is open at high temperatures and closes when the temperature drops below a certain level, turning on the heating system. When the temperature rises above the critical level, the switch opens again and the heating stops working.

The temperature-sensitive element is generally a bimetal strip (figure 6.26). The upper metal of the bimetal strip expands more than the lower when heated, and so

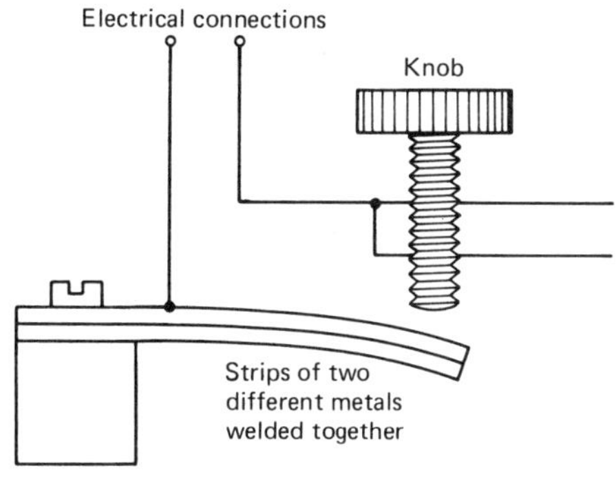


Fig. 6.26

the whole strip bends downwards when hot, and up again on cooling. The temperature at which the circuit is closed can be altered by adjusting the knob; screwing it down means the strip has to cool down less before touching it and so completing the circuit, and thus a higher temperature is maintained. As the heating system generally takes a little while to have any effect it is customary to include a small electric heating coil in the thermostat box. As soon as the contacts close this is switched on and stops the contacts coming open again if there is a cold draught before the main heaters get going (in electronic jargon this small heating coil adds hysteresis to the system.)

This kind of thermostat has the disadvantage that the main heater is either full on or completely off, and that there tend to be continual sudden small temperature fluctuations (to which biological systems are very sensitive), within the limits imposed by the thermostat. The situation is the same as that of a man trying to keep a room comfortable with an electric fire. He has to wait till it is getting too hot, then switch it off and wait till it is getting too cold before switching it on again.

Give the same man a gas fire from which the heat output is continuously adjustable and he can quickly find a suitable setting to produce a steady temperature and will need to make no further change till the external conditions change. I think that this sort of continuous control is obviously more suitable for biological work than the on–off systems. The first problem is to provide the large electric fan heaters normally used in constant-temperature rooms with a control analogous to a gas tap.

The best bet would seem to be a triac voltage regulator of the sort sold as power-tool speed regulators or lamp dimmers. These work by using a semiconductor switch to disconnect the mains supply from the electric drill/heater/lamp for part of each mains cycle (figure 6.27). The knob on the controller box changes the value of a variable resistor which controls the proportion of the time the lamp is connected to the mains, and hence how much power it consumes.

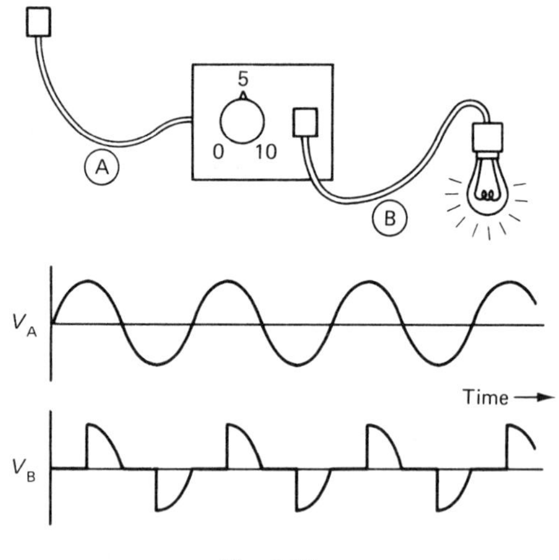


Fig. 6.27

If all the heaters in a constant-temperature room are controlled by such a device, a change in room temperature will cause the knob to be rotated by a proportional amount, thus altering the heaters' power. As a result, the temperature of the room will be well controlled and never subject to sudden fluctuations.

An excellent control system has been recently described; the details are outside the scope of this book, but they can be found in an article by R. M. Marston entitled 'Electric Heater Control', in *Wireless World* (June 1972).

## 6.3 MOVEMENT

### Activity Detectors

The simplest sort of movement detector gives only a measure of an animal's activity, that is, it tells you whether it is standing still or moving about. Four types of activity detector are commonly used, mechanical, capacitance, ultrasonic and microwave.

*Mechanical.* In this system the animal's movements alter the position of its cage, as shown in figure 6.28. If the animal leaves the centre of the cage one or other of the switches operates. Another arrangement is given in figure 6.29; as the insect buzzes

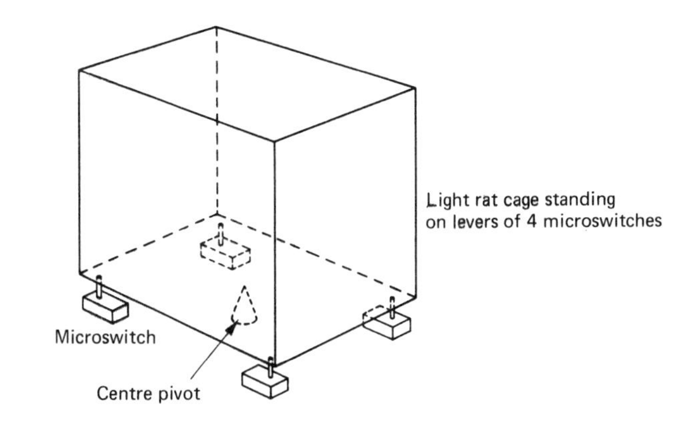


Fig. 6.28

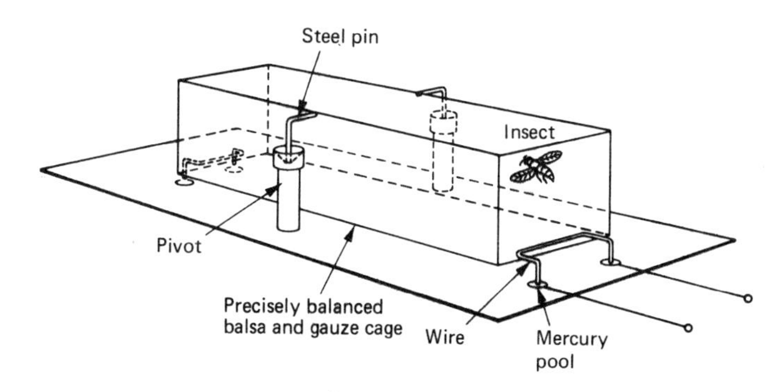


Fig. 6.29

from end to end the cage rocks, and the wire loops connect and disconnect the pools of mercury. The number of times per minute the switches operate is recorded on a printing counter, and gives a rough measure of the animal's activity levels.

*Capacitance.* If the insect in figure 6.30 walks onto a space above one of the metal strips, there will be a small change in the capacitance of the capacitor whose plates are the interleaved foil strips, for it will not longer have the insect as part of its dielectric. Given a really sensitive capacitance-measuring circuit, this device can thus be used as a movement detector. It will work well for anything except small insects, but does need careful screening from electrical interference.

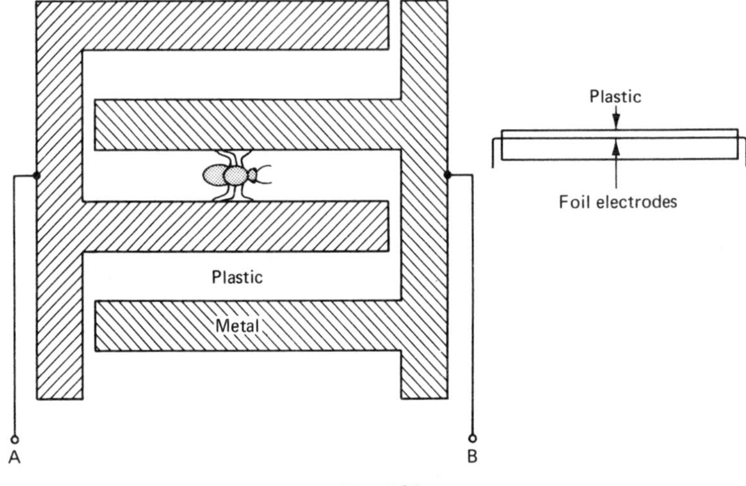


Fig. 6.30

*Ultrasonic and microwave.* These both involve putting the animal into a solid box containing an emitter and a receiver. The emitter sets up standing-wave patterns of either sound or electrical waves, and any disturbance of these caused by the animal moving shows as a change in the output from the receiver. The main disadvantage of both systems is a failure to differentiate between minute and gross movements of the animal. I have frequently seen an ultrasonic movement-detector give an enormous response when a rat merely twitched a whisker.

Of course it is necessary to be sure that the animal is wholly insensitive to the measuring device. Ultrasonic sounds can be detected by bats, crickets and some moths, for instance, and sub-millimetre wavelength electromagnetic radiation is indistinguishable from far infra-red light, and should be used with caution on animals (e.g. slugs, some moths) known to have infra-red sensors.

**Treadmill**

At the next level in sophistication in movement detectors are those which give some measure of the extent and direction of the movement. The simplest case is the treadmill (figure 6.31). A small protrusion on the rodent exercise wheel closes a microswitch each time the wheel goes round. The switch works an electromagnetic counter. Still more sophisticated types of treadmill are used to measure insect movement (figure 6.32). In these, microswitches and mechanically operated contacts

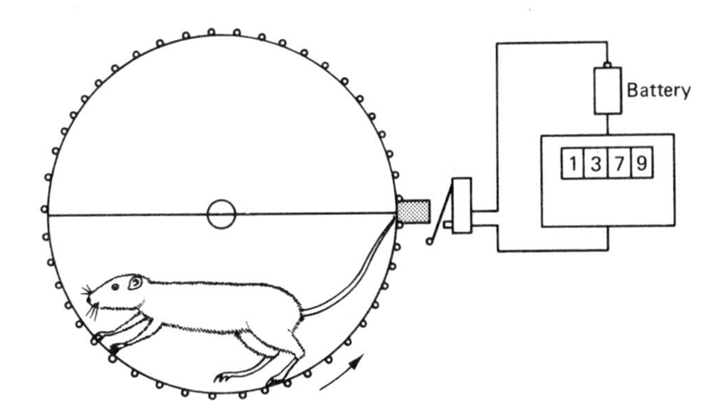


Fig. 6.31

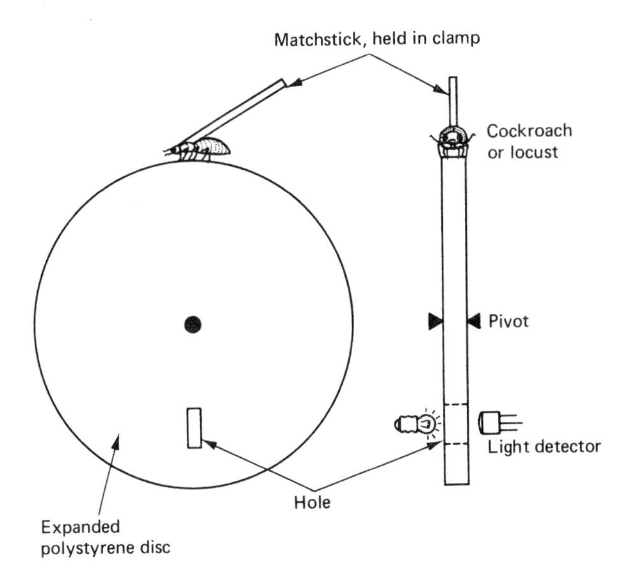


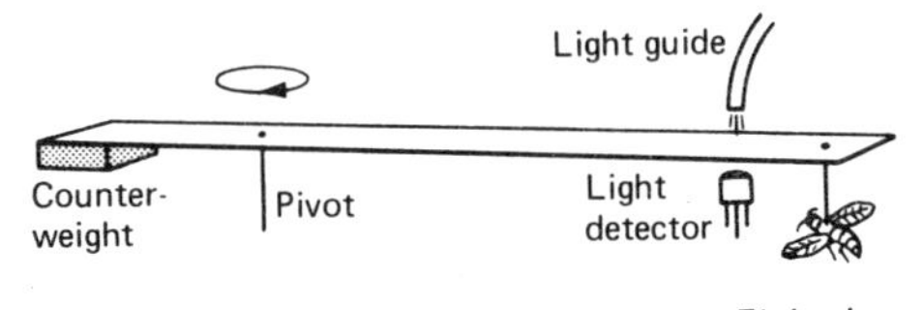


Fig. 6.32

are avoided, for if they are delicate enough not to impede the movement of the wheels noticeably they will almost certainly fail to make a reliable contact.

The use of an interrupted light-beam instead to detect movements of the wheel or bar has been facilitated by the recent introduction of light-operated integrated circuits. These function as comparators with a light level acting as input signal. With a light level below the threshold level the output voltage is zero volts. When the light level exceeds threshold, the output goes at once to +24 volts. The threshold level is set by the values of an external resistor and capacitor. These devices work at rates of up to 5 kHz (provided a reasonably bright light signal is available) and have sensitivities extending well into the infra-red. Their outputs will drive an electromagnetic counter via a single transistor switch (see chapter 5 and

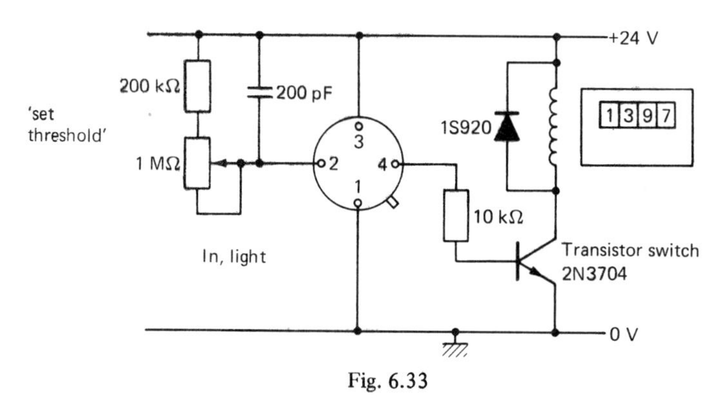


Fig. 6.33

figure 6.33). The light source for the integrated-circuit detector should be covered with a filter which passes only infra-red light if the animal's activity in the dark is to be observed.

## Position detectors

The treadmills give you the speed with which an animal moves in one constrained direction, but tell you nothing about the direction in which it would move if permitted to move freely. There are two approaches to this problem.

(a) the parallel channel approach, best exemplified by a cunning device designed by Richard Gregory (figure 6.34). A camera is set up on a tripod, and then pointed at, say, the entrance to an animal's burrow. A sharp image of the interesting areas is obtained on the ground glass screen by focusing the camera, and the photocells are put in strategic positions on this image by adjusting the magnets and clamps. It might be, for instance, that a dimming at photocell A followed quickly by a dimming at photocell B would mean that the animal had taken a certain route into its burrow.

Chapter 5 describes some electronic circuits which will enable this sort of event to be recognised and counted. This system has no limits to the number of separate photocells which are used, and, as they are all working all the time (a 'parallel channel'

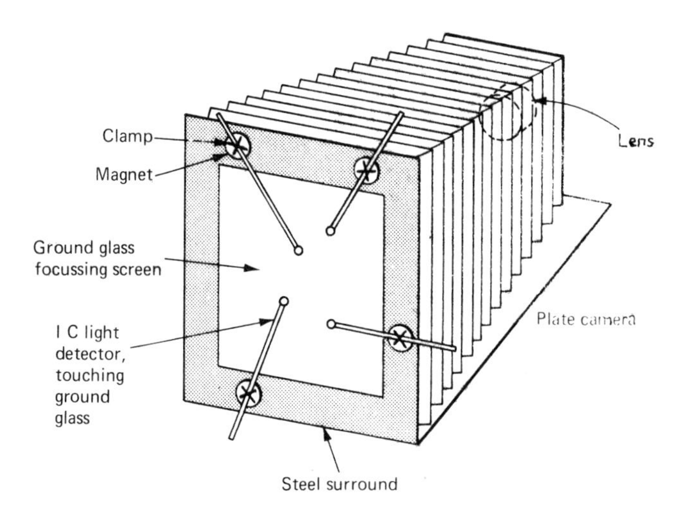


Fig. 6.34

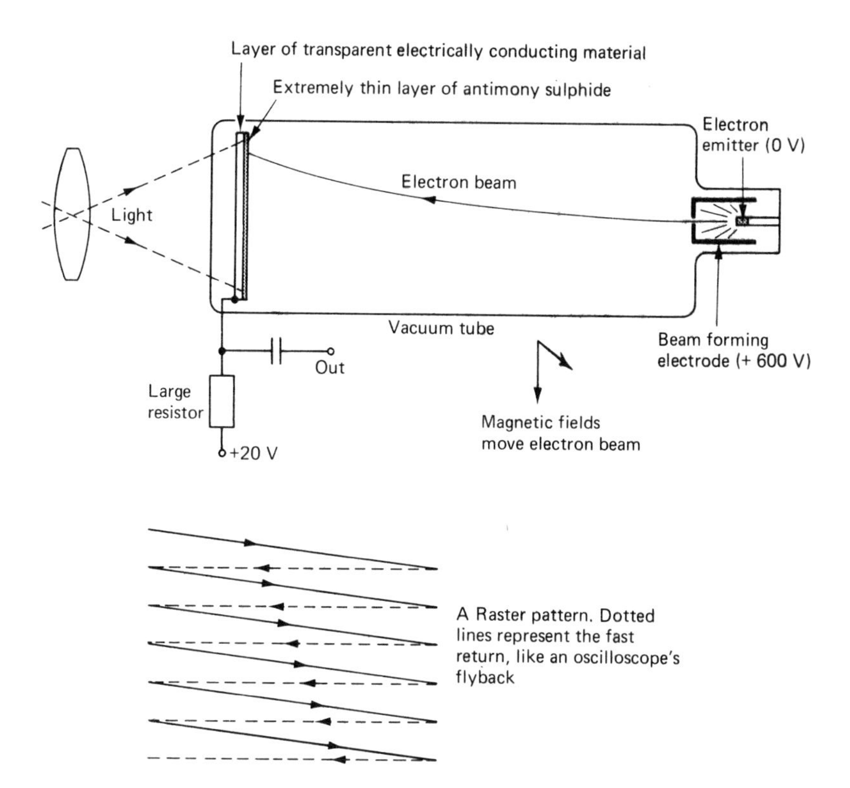


Fig. 6.35

system) the only limitation on the rate at which changes can be detected is the response speed of each photocell.

(b) Scanning systems. In these systems there is in effect only one photocell, and this is used to sample the light intensity of different regions of an image of the relevant area one after another. The spatial patterns of light and dark of the image are transformed into a temporal sequence of voltage changes as the one small light-sensitive area is moved around the image and samples every part of it in the course of time.

The vidicon (television camera tube) is a typical example of a scanning system (figure 6.35). A lens forms an image on a thin screen of light-sensitive material, antimony sulphide in this case, whose electrical resistance depends on the light intensity falling on it. This resistance can be measured for a tiny part of the screen by shooting an electron beam at it through the vacuum of the tube. We then have the situation shown in figure 6.36. The large resistor ensures that a constant current

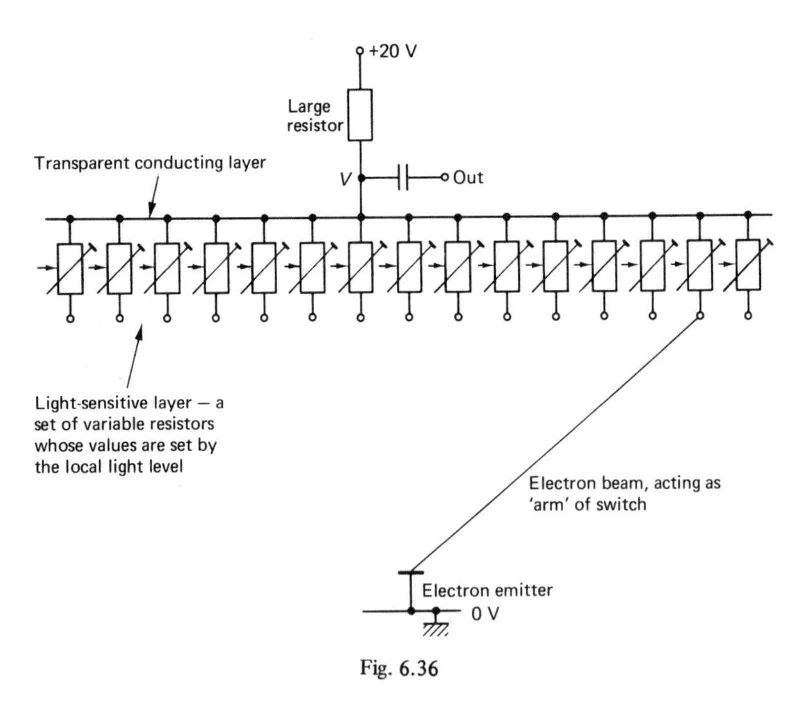


Fig. 6.36

flows through the light-sensitive layer and the electron beam to ground. Hence the voltage $V$ depends on the resistance of the light-sensitive layer at the place where the electron beam is hitting it, or in other words on the light intensity falling on the layer at this position.

Over the couse of 1/50 second the electron beam is moved over the image in a pattern of parallel lines called a raster by means of two electromagnets at right-angles to each other and placed just outside the vacuum tube. When it reaches the bottom right-hand corner of the picture the beam suddenly returns to the top left and begins to trace out another set of parallel lines, once again taking 1/50 of a second to cover the whole picture. This second set of lines falls in gaps between the first set, so the whole picture is covered (625 lines are sampled) in 1/25 second. The reasons for having two interlocking half-pictures is that when the picture is reconstructed with a modulated spot produced on a cathode ray tube phosphor by an electron beam moving exactly in step with that in the camera, an impression of flicker is avoided.

The movement of the spot in the raster pattern is called scanning, and is caused by suitable changes in the currents in the electromagnet coils. Figure 6.37(a) shows

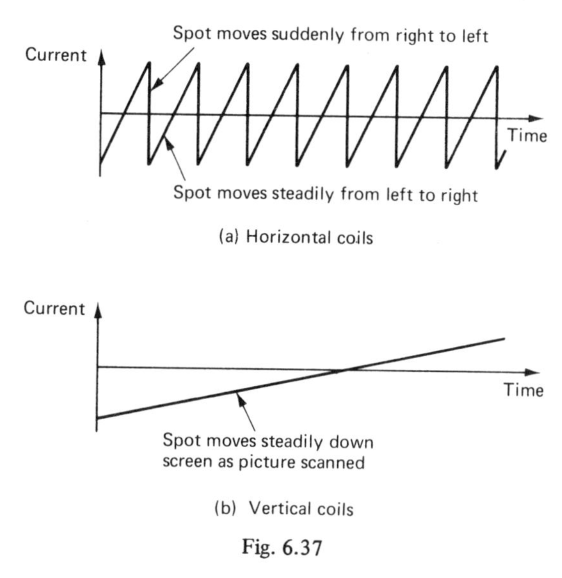


Fig. 6.37

the current changes for the magnet controlling the movement of the spot across the screen, and figure 6.37(b) for the magnet controlling the movement of the spot up and down the screen. In order to reproduce the picture a television receiver needs information about

(1) where the camera tube is sampling from
(2) what the brightness of the light is at that point.

Both sorts of information are transmitted in a single signal (figure 6.38).

In order to use a television camera to determine the position of a single bright object in the picture you generate a fast and a slow voltage ramp corresponding in shape to the current ramps shown above, and reset them to zero volts with the thin and fat sync pulses respectively (the sync pulses can be separated from the brightness signals with a simple diode circuit, and the fat ones separated from the thin ones by a smoothing circuit).

At any instant the value of the slow-ramp voltage is proportional to the $Y$-axis of the spot on the picture monitor, and that of the fast ramp to the $X$-axis. The brightness level signals are monitored by a comparator, and when they exceed a certain level (i.e. the electron beam in the camera reaches the bright object) the comparator goes over and makes two sample-and-hold circuits (see chapter 4), one looking at each of our ramps, sample briefly and hold the ramp voltages. The sample-and-hold outputs give a continual record of the $X$ and $Y$ coordinates of the position of the bright object in the field, and these values are updated each time a picture is

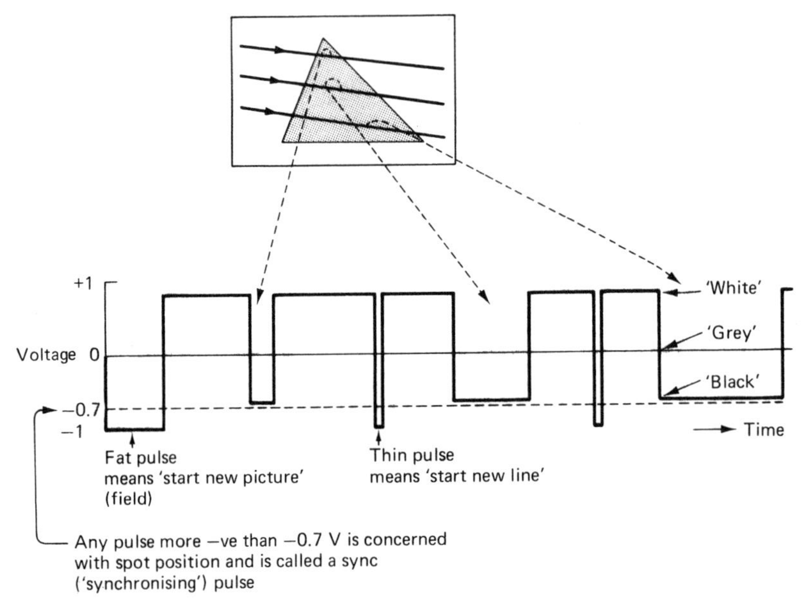


Fig. 6.38

scanned. This sort of system clearly removes an enormous amount of labour from experiments involving tracking single animals.

It is possible to get television camera tubes which work well in dim light (called plumbicons) and vidicons which are infra-red sensitive. It is not generally possible to speed up or slow down the scanning rates of vidicons appreciably, nor is it easy to see how to cope with more than one object at once.

A poor man's version of this machine can be made with an oscilloscope. A raster is generated on the screen by using the internal timebase to produce each line, and a slower ramp input to the $Y$-amplifier to move the 'line' steadily down the tube face. The oscilloscope is stood on its back, tube pointing up in the air, and the slow-moving insect to be tracked placed on the tube face. When the spot passes beneath the insect a watching photocell tells two sample-and-hold circuits to sample the timebase ramp and the one driving the $Y$-amplifier. Of course the insect need not be on the tube face; instead a lens can be used to project an image of the moving spot onto the arena in which the insect is wandering about, and a very sensitive photocell can gather the reflected light from the whole area.

This system can run at any speed you like, but cannot be made to work with infra-red light because there are no infra-red phosphors. There are, however, phosphors giving a good red light, to which some animals are insensitive.

### Movements of Parts of Animals

The currently favoured system for monitoring limb and appendage movements in unrestrained animals is to use a Hall-effect generator. This is a thin square piece of the semiconductor indium antimonide, with electrical connections to the midpoint of each side of the square. A steady current is passed between two opposite connectors and the voltage between the other pair is measured. This voltage is directly proportional to the strength of the magnetic field impinging on the semiconductor. A steady magnetic field strength produces a steady output voltage, in contrast to the situation in which one attempts to monitor a magnetic field by measuring the current induced in a coil of wire, when only changing fields produce any output voltage.

The Hall-effect generator is fastened to the body of the animal, and a small magnet to the moving appendage. The output from the Hall generator thus gives the appendage position. An elegant application of this technique on a micro-miniature scale is Elsner's work recording the limb movements of a stridulating grasshopper. (N. Elsner, 'The recording of the stridulating movements in the grasshopper *Chorthippus mollis* using Hall-Generators', *Z. vergl. Physiologie,* **68**, 417–428 (1970); a German paper with an English summary.)

## 6.4 HUMIDITY

The only really reliable humidity-measuring devices consist of a piece of hair under tension, whose length changes as it absorbs and loses water. The moving end of the hair works a pointer which moves over a scale calibrated in relative humidity (the ratio of the actual water content of the air to what it would contain under the same conditions if it were saturated). Hair hygrometers, as they are called, are accurate to about 5 per cent, and can be made reasonably small. If one switch contact is placed on the scale and another on the pointer the device becomes a humidistat and can control a humidifier to produce an environment of constant humidity. Humidifiers consist either of an electric kettle to pass steam into a stream of air, or of an aerosol water spray. The kettle systems tend to go steaming long after they are switched off, and control is rarely to more than ± 5 per cent relative humidity. The spray systems, with very careful attention to air flow, can give control to ± 2 per cent.

To improve on this sort of control or to measure the humidity in a really small space is quite ridiculously difficult. The most promising method of humidity detection for a microhygrometer seems to me to be dewpoint measurement. This involves cooling a piece of metal in the atmosphere whose humidity is to be measured (at say 20 °C) and noting the temperature at which dew begins to form, say 5 °C.

This tells you that there is as much water in the air as there is in water-saturated air at 5 °C. Tables tell you how much water vapour there is in water saturated air at 5 °C and at 20 °C, and the ratio between these values gives you the relative humidity.

In my current prototype the mirror is cooled by passing an electric current through a frigistor (a thermocouple working backwards), and a photocell stops the cooling and causes the dewpoint temperature to be measured when dew forms on the mirror. This approach seems more promising than measuring the electrical resistance of bits of polystyrene or the temperature difference between wet and dry thermistors.

## 6.5 ELECTRICAL RESPONSES

The best tool for recording the electrical signals from inside a nerve cell is the glass micropipette, a glass capillary tube drawn to a minute (0.5–1 μ diameter) point and filled with salt solution (classically three molar KCl, but some entomologists use Ringer, a much more dilute solution of ionic concentration corresponding to the blood of the animal concerned) or with a low melting point alloy of Wood's metal and a little indium.

The mysteries of electrode pulling, filling and the like are difficult to learn from books – you need to become apprenticed to a practising electrophysiologist and have the skills passed on to you. If this is not possible one can fortunately forgo these mysteries and buy ready-made electrodes.

The electronics involved in electrophysiology is straight-forward. A DC-coupled high input impedance preamplifier (see Chapter 4) is essential.

Signals from the preamplifier go to an oscilloscope, a tape recorder and an audio monitor (a box containing an audio amplifier, which is available on an integrated-circuit, and a loudspeaker).

Data analysis is best done on the tape-recorded signals after the experiment is over. If a tape-recorder channel can be spared it is an excellent idea to have a microphone to record the experimenter's comments.

# 7 Automated Experiments

Many biological experiments require a complicated test to be applied to a large number of subjects or experimental animals if a significant trend is to be discovered. This makes life hard for the experimenter, who is faced with many weeks of very repetitious work. Moreover it is unlikely that the experimenter will be able to apply the test in a completely stereotyped fashion each time, and thus an additional source of variability will be introduced into the results. In this kind of situation it is worth trying to arrange for completely automatic stimulus presentation and data collection and analysis. One should allow on average at least two to three months development time for a completely automated apparatus, so this sort of project only pays off if the experiment is going to be very lengthy indeed, or else is a standard assay technique which is to form the basis of a research programme.

It is all too easy to spend a month making a complex apparatus to perform an experiment which could have been done with a fortnight's concentrated observation by a human observer equipped with paper, pencil and stopwatch. Automated set-ups replace the experimenter's stopwatch with a process timer, and his hands with electric motors.

## 7.1 BUILDING BLOCKS FOR AUTOMATED EXPERIMENTS

### The Process Timer

Switches in which everything happens inside a semiconductor block (e.g. transistors, triacs) are much superior to those with moving metal contacts because they neither wear out nor cause sparks, which often interfere with other apparatus. However they cannot generally be used to switch the AC mains supply to pieces of apparatus, for reasons of expense, inconvenient electronics, and frequently, fire regulations which forbid any electronics to remain connected to the mains when an appliance is switched off. Hence if, for instance, it is necessary to switch a tape recorder on for five-minute periods once an hour through the night the best sort of switch is probably a mechanical one worked by a clock motor. These are called process timers, and are very cheap because they are made in huge quantities for automated factories (figure 7.1). The interval to be timed is selected by turning the knob till the upper pointer is over the appropriate number on the scale. So long as the clutch is not

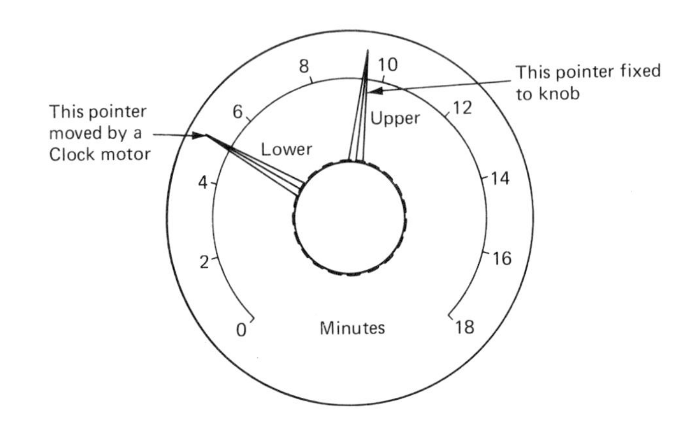


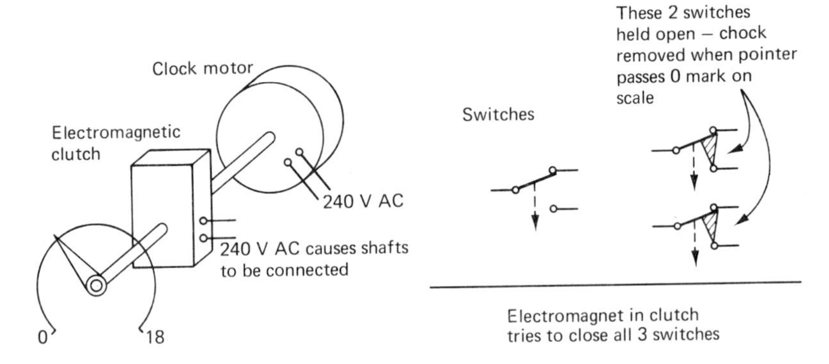


Fig. 7.1

engaged the two pointers are both moved by the knob and stay exactly together, for a spring pulls the lower pointer anticlockwise round the dial and holds it firmly against a downward projection from the upper pointer.

To start the timer, power is supplied to the motor and the clutch, and then the lower pointer begins to move towards the zero minutes mark, against the pull of its spring. The clutch electromagnet causes one of the internal switches to close as soon as the clutch is engaged. The other two switches try to close too but are prevented by blocks of plastic. If the power supply to the clutch is cut off at any stage the lower pointer springs back to its starting place under the upper pointer.

When the lower pointer reaches the zero mark it removes the plastic blocks and the remaining two switches can be pulled down by the clutch electromagnet. Switching off the clutch lets the lower pointer and all three switches return to their initial positions, and reinserts the plastic blocks.

All three switches will handle 240 volts AC, and so can be used to control the clutch or the motor power supplies. The timer is supplied with one of the switches internally connected to the clutch and motor in a most puzzling fashion. These connections are easily altered, and it is best to start thinking about each application without any preconceptions about which switch should be connected to what.

(*i*) *To output a brief pulse every ten minutes* (figure 7.2). When first switched on a current flows down wire A, through switch 1, and magnetises the clutch. This starts the pointer moving, and pulls switch one into the down position so the clutch coil

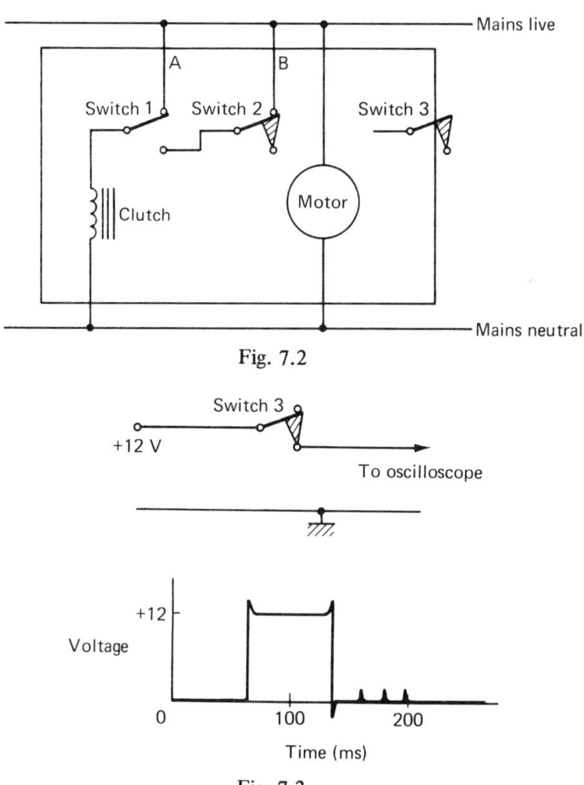


Fig. 7.2

Fig. 7.3

now gets its current down wire B and through switch 2. Ten minutes later, when the pointer reaches time zero on the scale, the chock is pulled from switch 2, and the clutch magnet can pull the switch down, thus interrupting its own power supply. The pointer is disconnected from the motor and springs back to the '10 minute' calibration mark, and both switches return to the up position. Current now flows down wire A to remagnetise the coil via switch 1, and the whole cycle recommences.

The output is obtained in the form of the brief change in switch 3's connections each time the system resets. The only snag with this sort of system is that the contacts do tend to bounce a little during the reset period; figure 7.3 shows what the actual output tends to look like. Care must be taken to ensure that whatever the timer is controlling responds to the initial switch closure, and not the subsequent little spikes. This problem is acute if the device the timer is switching on has electronic circuits, for it is difficult to get electronic and electromechanical systems to work well together. The best solution may well be to get the first positive-going voltage change to trigger a delay circuit, say of 500 milliseconds duration, and start the rest of the circuitry operating when the delay-circuit output returns to zero volts, long after the timer has sorted itself out.

(*ii*) *To switch on an apparatus for 30 seconds once an hour* (figure 7.4). The push switch starts the 1 hour clock, which, at the end of its period, starts the 30 second

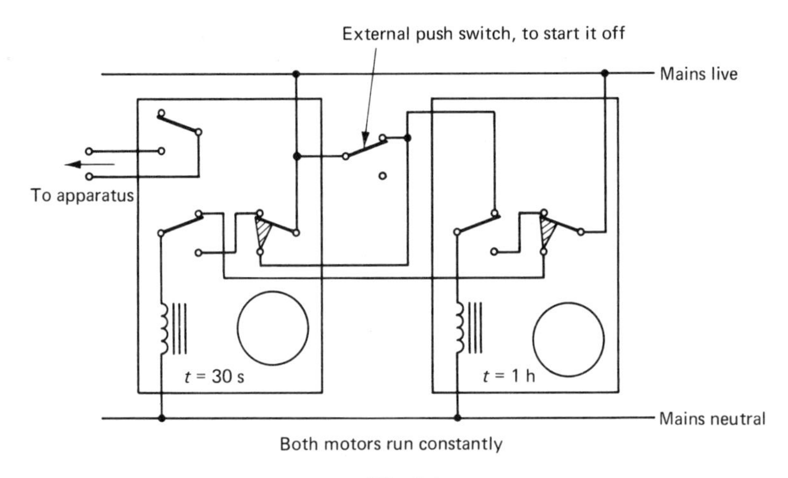


Fig. 7.4

clock, and resets. It does not, however, commence timing again until the 30 second clock reaches the end of its period, when the cycle recommences. The extra set of contacts on the 30 second clock works the apparatus concerned.

## Electric Motors which Rotate at Constant Speeds

These are called synchronous motors because their motion is controlled by the 50 cycles per second oscillation of the AC mains. They are commonly used in electric clocks. The principle is straight-forward: a magnet is pivoted so that it can rotate inside a coil (figure 7.5). A motor of this sort is incapable of starting by itself. If you switch on the power the magnet just rocks slightly back and forth 100 times a second. But if you spin it by hand so that it reaches a speed of 50 rotations per second the changing magnetic field will lock on to it and the self-perpetuating system shown in the figure will take over. The direction of shaft rotation depends on the direction of the initial spin.

Synchronous motors which go slower than 3000 rpm (50 revolutions per second) have a rotating magnet with more than two poles. Synchronous motors sometimes have built-in mechanisms for getting them up to the speed at which the synchronising mechanisms takes over, generally in the form of an auxiliary motor whose speed

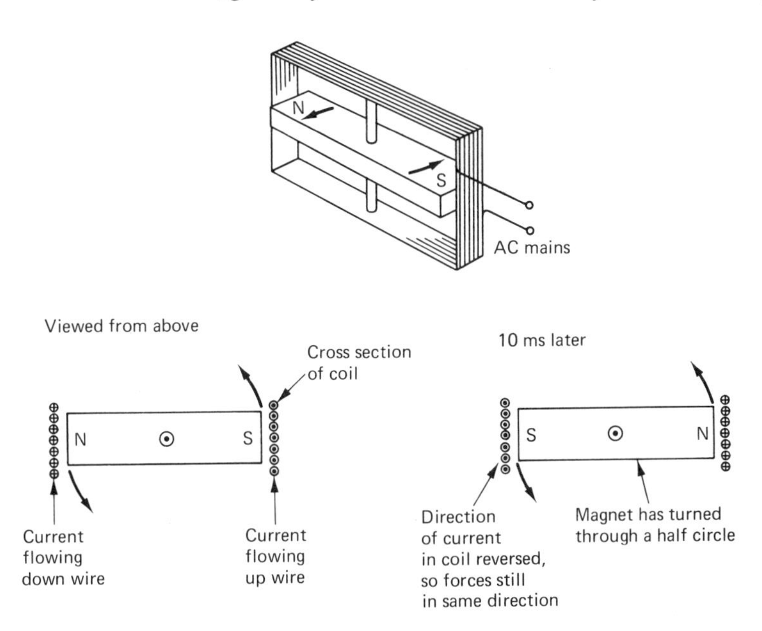


Fig. 7.5

is not controlled by the mains frequency. If this is the case they will always start in the same direction.

Versions with built-in gears, so that the output shaft rotates once a minute, or even once an hour, are available. In general it is always a good idea to look for a motor which goes at the correct speed for your application, for making gear chains yourself calls for a very high degree of engineering skill. Synchronous motors are the obvious choice for moving visual displays at constant speed and similar applications calling for a very low-power motor.

### Stepping Motors

Stepping motors are a cunning version of the synchronous motor in which the shaft is locked solid in one position till a voltage pulse is input to the system, whereupon it jerks round through a precisely controlled angle of a few degrees, then locks solid again. A continuing chain of input pulses produces a constant speed of rotation, and if necessary it is easy to smooth out the little jerks by connecting the motor

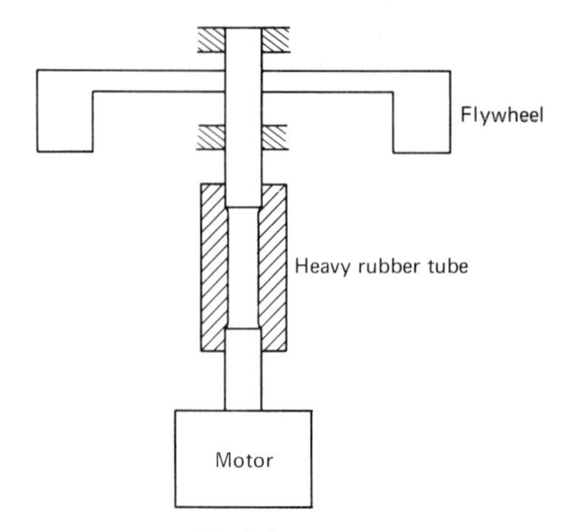
Flywheel
Heavy rubber tube
Motor

Fig. 7.6

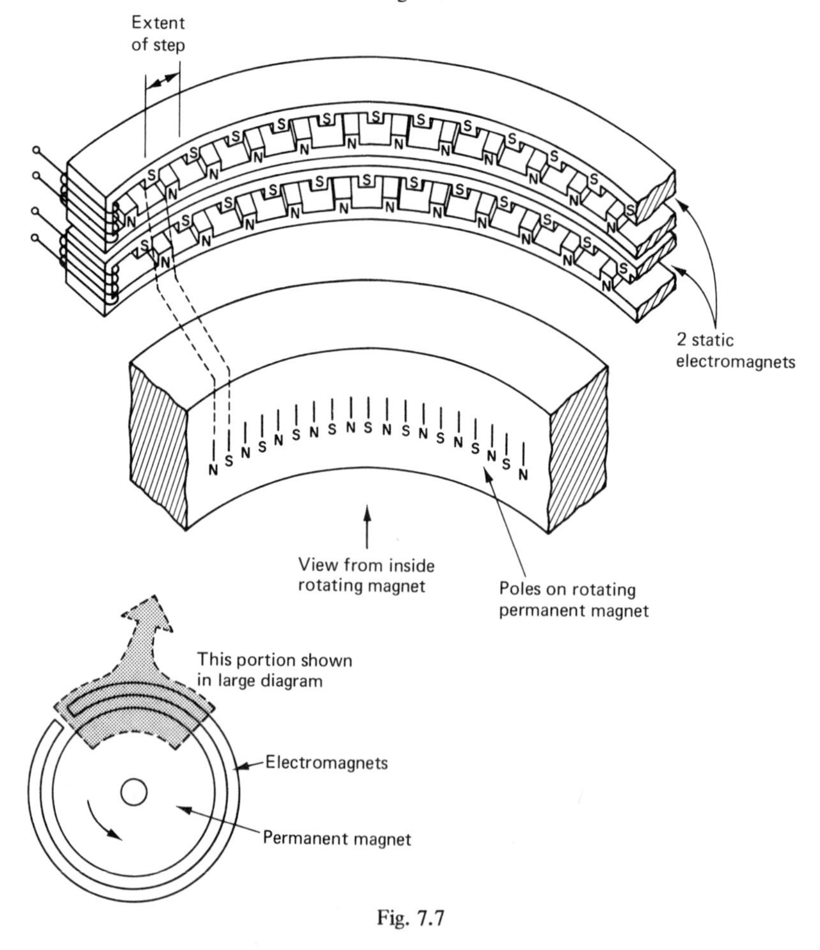
Extent
of step
2 static
electromagnets
View from inside
rotating magnet
Poles on rotating
permanent magnet
This portion shown
in large diagram
Electromagnets
Permanent magnet

Fig. 7.7

shaft to a heavy flywheel via a flexible coupling, for example a length of heavy-duty rubber tubing (figure 7.6). Speeds up to 200 revolutions per minute are possible.

Stepping motors have two stationary electromagnets and a moving permanent magnet (figure 7.7). If the current is held constant in both of the electromagnet coils the arrangement of north and south poles will be as shown, and the permanent magnet rotor will be held firmly in one position with each of its south poles between two equally attractive north poles, one on each electromagnet; each of its north poles will be similarly fixed between two electromagnet south poles.

If the connections to the top electromagnet are now reversed, causing the positions of its north and south poles to interchange, the rotor finds itself in a highly unstable position with each of its poles being repelled by a similar pole on the right and attracted by a dissimilar one on the left. So it quickly jumps one step to the left (anticlockwise) and the *status quo* reasserts itself. To make it take another anticlockwise step you now have to change over the connections on the lower electromagnet.

To make the changing of the direction of the current easier it is normal to wind two coils on each electromagnet, as shown in figure 7.8. Moving the switch reverses

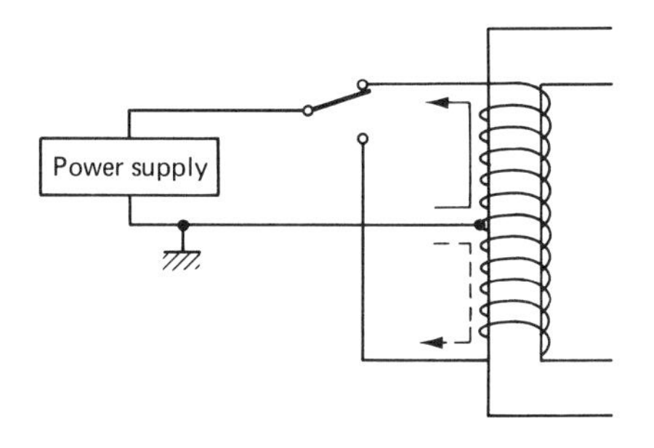


Fig. 7.8

the current direction. The switch generally takes the form of two transistors, and to drive the four transistors controlling the two electromagnets one uses bistable memory circuits (figure 7.9).

If one output is at + 12 volts the other has to be at zero volts and vice versa. The behaviour of this circuit is described in chapter 5, where it is compared to a retractable ballpoint pen: one push, and the tip emerges, another, and it retracts. Bistable circuits can be bought ready-made as integrated circuits, or made up from two transistors and a few other components. Circuit details are given in chapter 8.

The arrangement required for a stepping motor is given in figure 7.10. When bistables 2 or 3 receive an input pulse they change over whichever of their two switch transistors is switched on, and thus the half of the electromagnet coil through which current flows. Each state change of bistable one is a valid negative-going input pulse for either bistable 2 or for bistable 3, and, as can be seen from the asterisks marking the relevant downward pulses on figure 7.9, a continuing chain of input pulses to bistable 1 causes bistables 2 and 3 to be triggered alternately. So the first

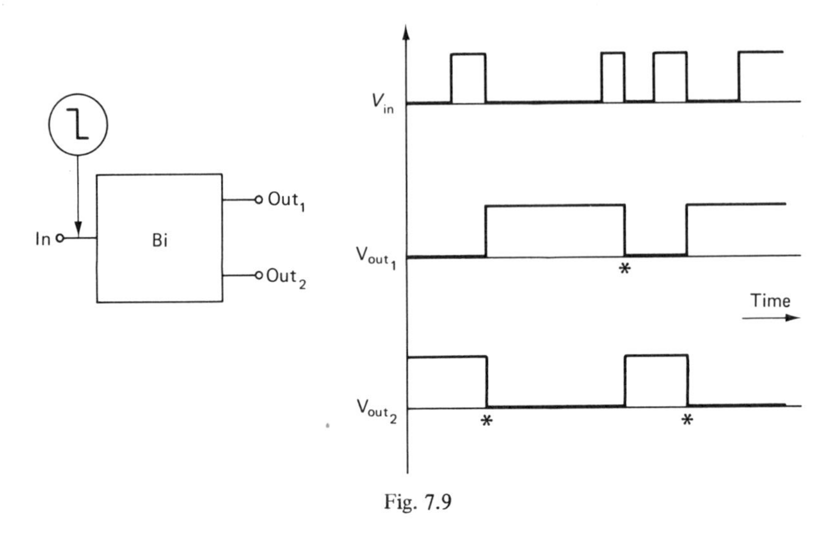


Fig. 7.9

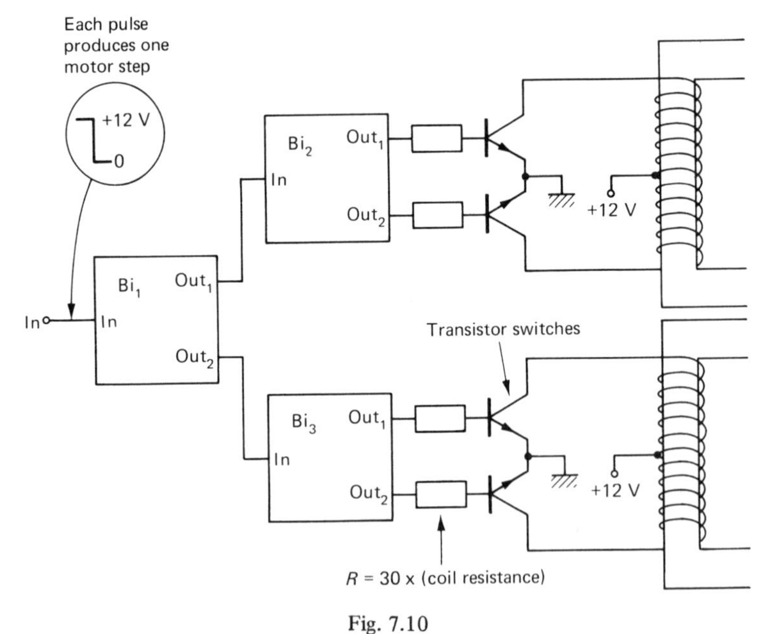


Fig. 7.10

input pulse changes the polarity of the top electromagnet, the second input pulse that of the bottom electromagnet, and so on, each producing a motor shaft rotation step in the same direction. The actual direction of rotation depends on the initial states of the bistables when the circuit is switched on.

Integrated-circuit bistables have a special 'reset' input, which, when given a brief pulse, turns them to a state with output 1 at zero volts. If all these reset connections are given a pulse immediately after switching on, thereafter input pulses will always produce the same direction of rotation. If you want the motor to turn the other way, just swap over the input connections of bistables 2 and 3.

**Velodynes**

Velodynes are motors capable of accurately controlled fast rotation. They involve a DC motor and DC generator mounted on the same shaft. DC motors are the familiar small electric motors with a static electromagnet pushing a moving one round in circles. The moving electromagnet gets its power via carbon brushes bearing on brass contacts (figure 7.11). To make the motion less jerky most DC motors have

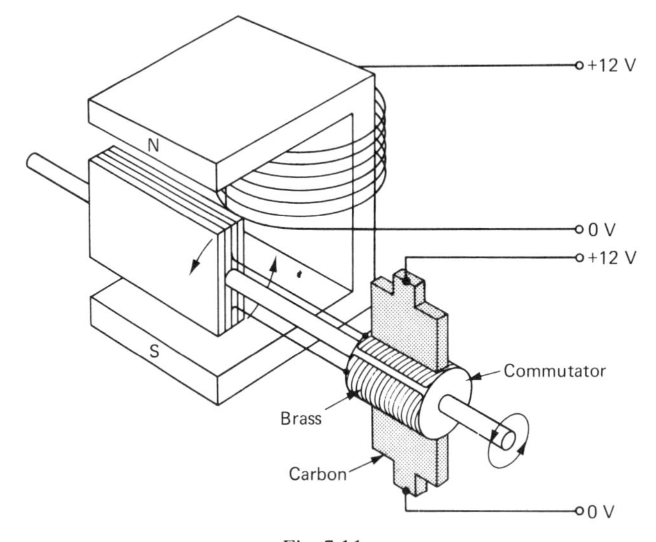


Fig. 7.11

more than one moving coil. Each coil has its own pair of commutator contacts, and is only connected up when it is in the position of the single coil in the diagram, that is when most force can be exerted upon it. If the static and moving electromagnets are connected in parallel the motor is called 'shunt wound' and tends to a constant speed. If they are connected in series it is called 'series wound' and tends to go as fast as it can, but is better at starting up under a heavy load. Series-wound DC motors also work on AC. Both magnetic fields change direction exactly in step with each other, so there is little nett effect on the motor's behaviour.

DC generators are DC motors pushing electricity out of their coil leads when you rotate the shaft. The static electromagnet needs either to be replaced with a permanent magnet, or to have arrangements made to pass some of the current from the carbon brushes through it. This system has to pull itself up by its own bootstraps, because you need some static magnetic field to generate any current to work the electromagnetic causing the static field, but there is generally enough residual permanent magnetism to get things going. Careful mechanical design can produce a DC generator whose output voltage is directly proportional to the speed of rotation of its shaft.

The velodyne uses such a generator, and compares its output to a pre-selected reference voltage. The motor's static electromagnet is driven by an amplifier whose input is the difference between the generator output and the reference voltage. If the motor deviates from a constant speed the generator voltages differs from the reference voltage and the amplifier takes appropriate correcting action. The constant speed produced is directly proportional to the reference voltage, and the direction of rotation depends on its sign (figure 7.12).

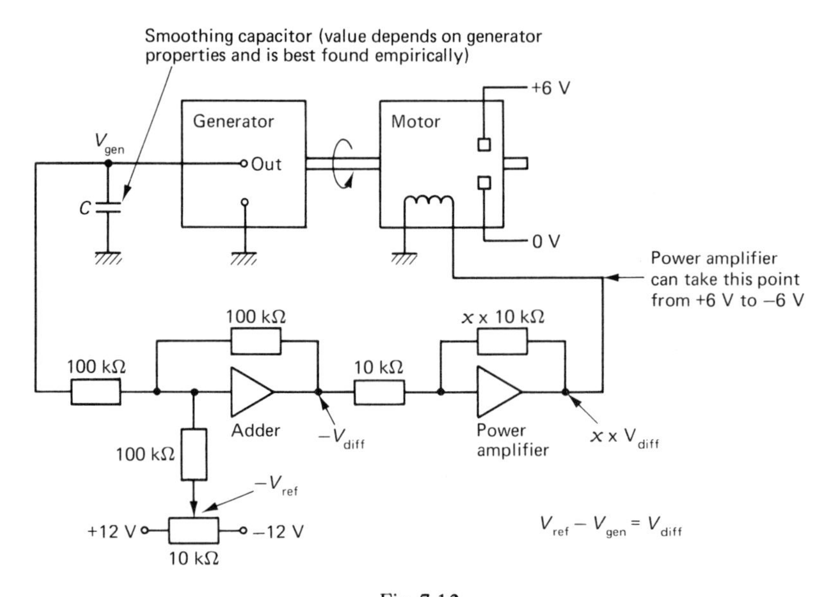


Fig. 7.12

The first operational amplifier finds the difference between reference and generator voltages by adding $-V_{ref}$ to the generator voltage. The second operational amplifier is a power amplifier capable of driving the motor's static electromagnet coil. The amplitude of the voltage across the coil controls motor speed, and its sign controls motor direction of rotation. The gain of this power amplifier is adjusted till the control system works well. If the gain is too low the correcting mechanism will be sluggish; if the gain is too high it will be over-vigorous and the speed will tend to oscillate as each correcting signal causes an overcorrection.

The simplest form of power amplifier is an integrated-circuit operational amplifier with a single power transistor to increase its current output (figure 7.13). The power transistor is connected as an emitter-follower (figure 7.14). This is a circuit in which the output voltage remains almost exactly equal to the input voltage ('non-inverting amplifier of unit gain'). It is useful because it has high input impedance and

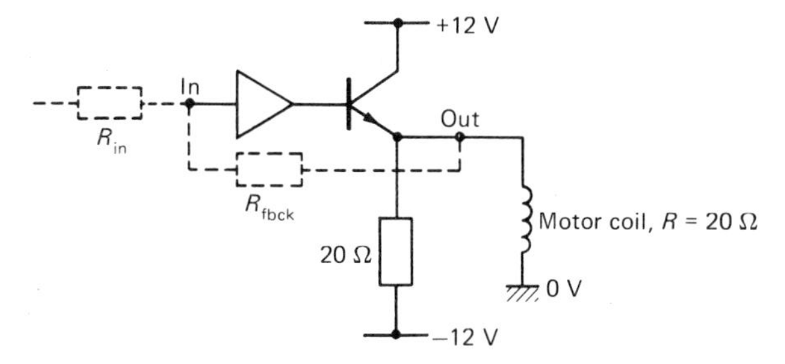


Fig. 7.13

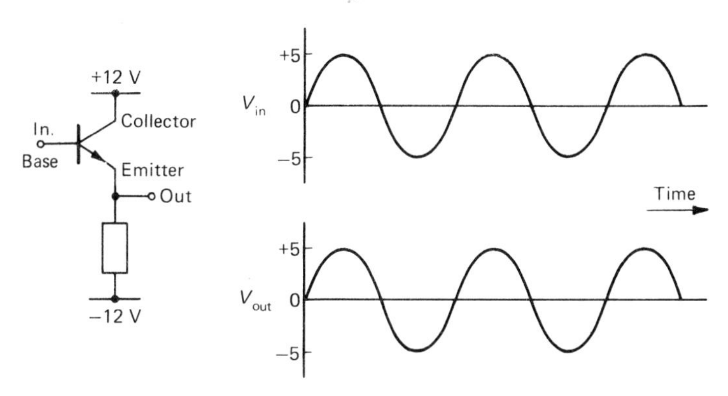


Fig. 7.14

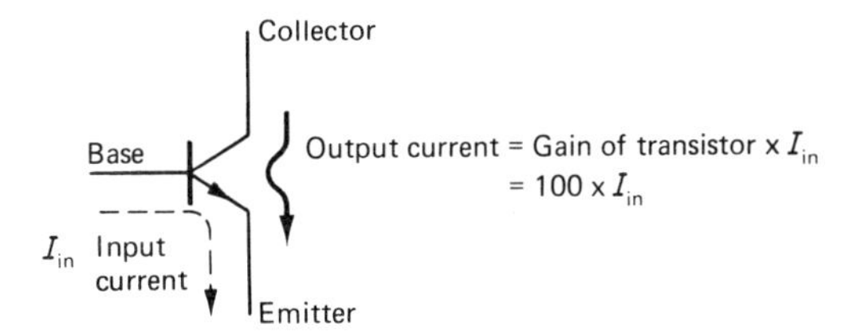


Fig. 7.15

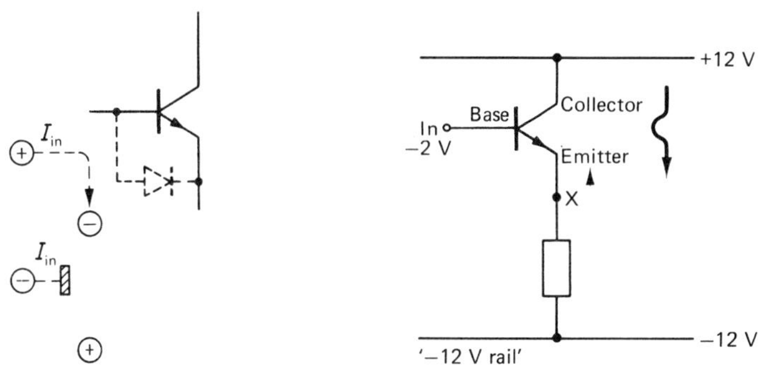


Fig. 7.16

Fig. 7.17

a low output impedance. Input currents are hence small, whereas output currents can be quite large if the load is of low resistance.

In any transistor the output current, flowing from collector to emitter, is around a hundred times the input current, flowing from base to emitter (figure 7.15). No input current means no output current. The conditions when an input current flows are determined by the voltage of the base relative to the emitter. Transistors behave as if there were a diode connecting base and emitter inside the transistor (figure 7.16). If the base is negative with respect to the emitter no input current can flow because the internal diode's high backward resistance prevents it. If the base is positive with respect to the emitter input currents can flow, and, because of the diode's low forward resistance, the base is effectively connected directly to the emitter, and the voltage at the base always equals the voltage at the emitter (plus the usual 0.5 volt across conducting diodes).

Consider an emitter-follower circuit (figure 7.17). If there is no connection to the input, point X is at − 12 volts because the transistor passes no current. If the input is connected to a point at, say, − 2 volts, the input wire (transistor base) will momentarily be 10 volts positive with respect to the emitter (point X) so a current will flow into the transistor, causing a much larger output current to flow from collector to emitter and down to the − 12 volt rail through the resistor. As the current flowing through a resistor increases so does the voltage across it (Ohm's Law), so point X moves rapidly away from − 12 volts towards + 12 volts. As soon as it passes − 2 volts the emitter becomes more positive than the base and the input current stops. The output current stops too, so the voltage at X drops towards − 12 again. As soon as it becomes more negative than − 2 volts the transistor input current recommences and pushes it up again. Clearly this is an example of negative feedback, X being maintained at the voltage of the input wire, whatever it may be, and thus holding the input current very low and supplying whatever current is necessary to maintain the input voltage across the load.

The case of our velodyne motor is a little difficult because we want to be able to reverse the current through it to make it go backwards. Hence the circuit in figure 7.18(a) will not suffice and we have to resort to that shown in figure 7.18(b). For inputs between zero and +12 volts this is straightforward (figure 7.18(c)). $I_R$, the current through the resistor, is wasted, but the circuit can work just as described above.

For negative input voltages things become more complex (figure 7.19). If no input connection is made to the transistor, point X will stay at a voltage determined by the coil and resistor $R$, which is acting as a voltage divider. If $R = R_{coil}$, for instance, X is at − 6 volts. Hence for all input voltages between − 6 and − 12 the transistor internal diode will be biassed backwards and no input current will flow. The smaller $R$ is made, the nearer the cut-off is to − 12 volts, and the more current is wasted when the input voltage is positive. To take a concrete example: if $R_{coil}$ = 10 ohms and $V_{coil}$ must be capable of being changed from + 4 to − 4 volts, then $R$ will equal 20 ohms. The largest current will pass through the transistor at $V$ = + 4 (figure 7.20). Hence a transistor capable of handling 2 amps and 10 watts would be safe for this application.

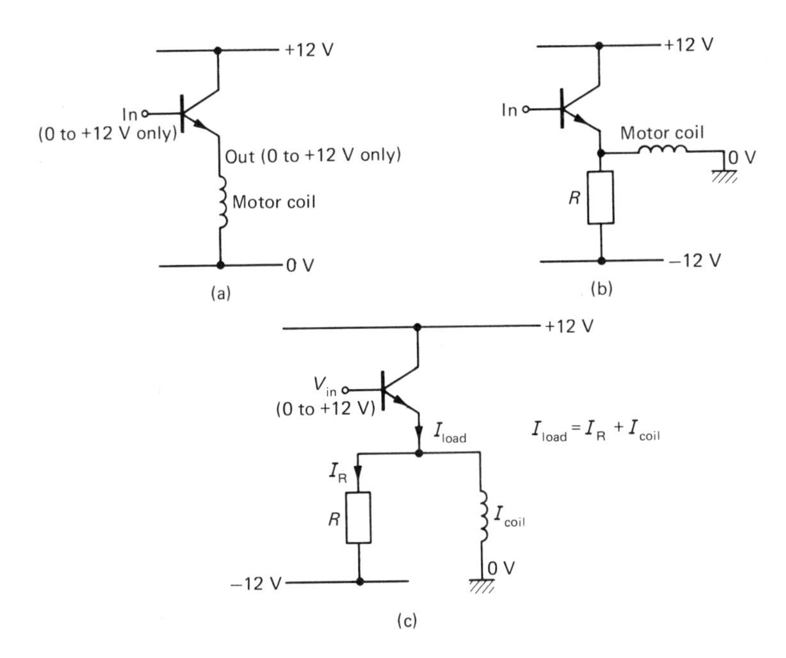


Fig. 7.18

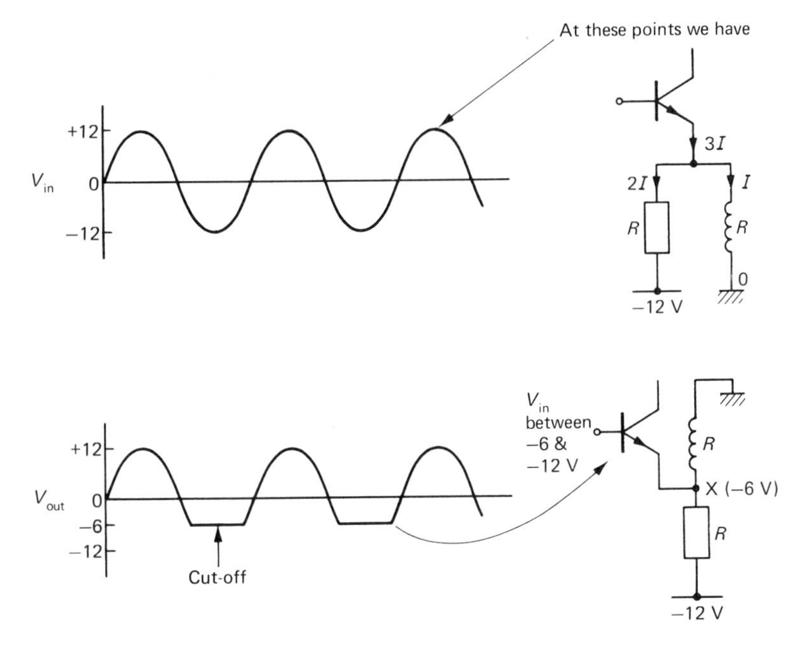


Fig. 7.19

The power supply must be capable of providing enough current, and the 20 Ω resistor must be sturdy enough to dissipate 15 watts of heat caused by the maximum current flowing through it. It is advisable to set the power supply overload protection to a current level below the maximum which the transistor can handle.

Alternative circuits which do not waste power on this scale involve using two power transistors, one of the sort shown, an NPN transistor, which works between the + 12 rail and ground, and one a PNP transistor, which is a type designed to work

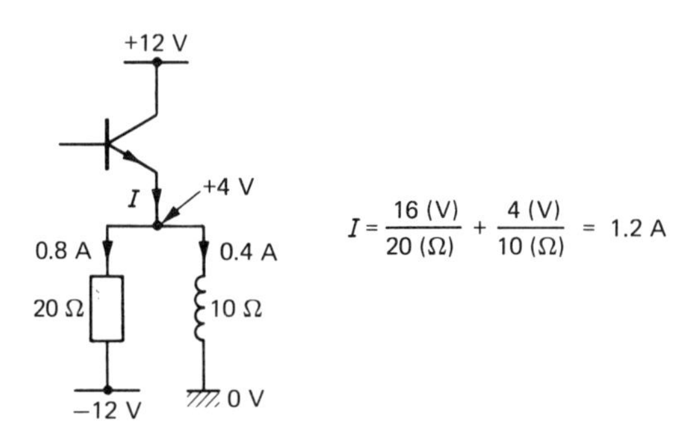


Fig. 7.20

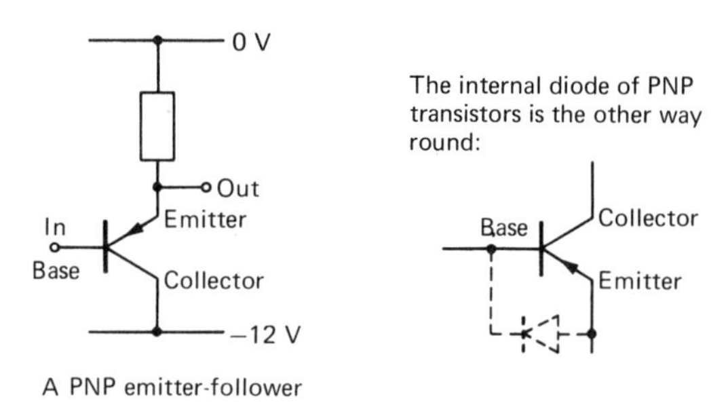


Fig. 7.21

between − 12 and ground (figure 7.21). Arranging things so that the two power transistors do not fight each other when the input voltage is close to zero takes considerable design skill, and even when these circuits are working well the 'join' between the jurisdictions of the two transistors tends to show.

### Some Practical Points about Motors

Velodyne motor-generators are expensive (at least £120) and very carefully designed so that the motor runs fairly well at the slower speeds. Most motors with commutators work badly below 2000 revolutions per minute; their speed is irregular and they have little torque. Hence velodynes made of any two old electric motors with shafts linked (in the pious hope that one of them is going to be a good generator) are not likely to work well at low speeds.

A better system for a home-made motor-speed monitor is to attach a transparent sectored disc (figure 7.22) to the motor shaft and to use a lamp and photocell to count the sectors. A suitable circuit for converting the chain of pulses output by the integrated-circuit light detector into a voltage proportional to the motor speed is given

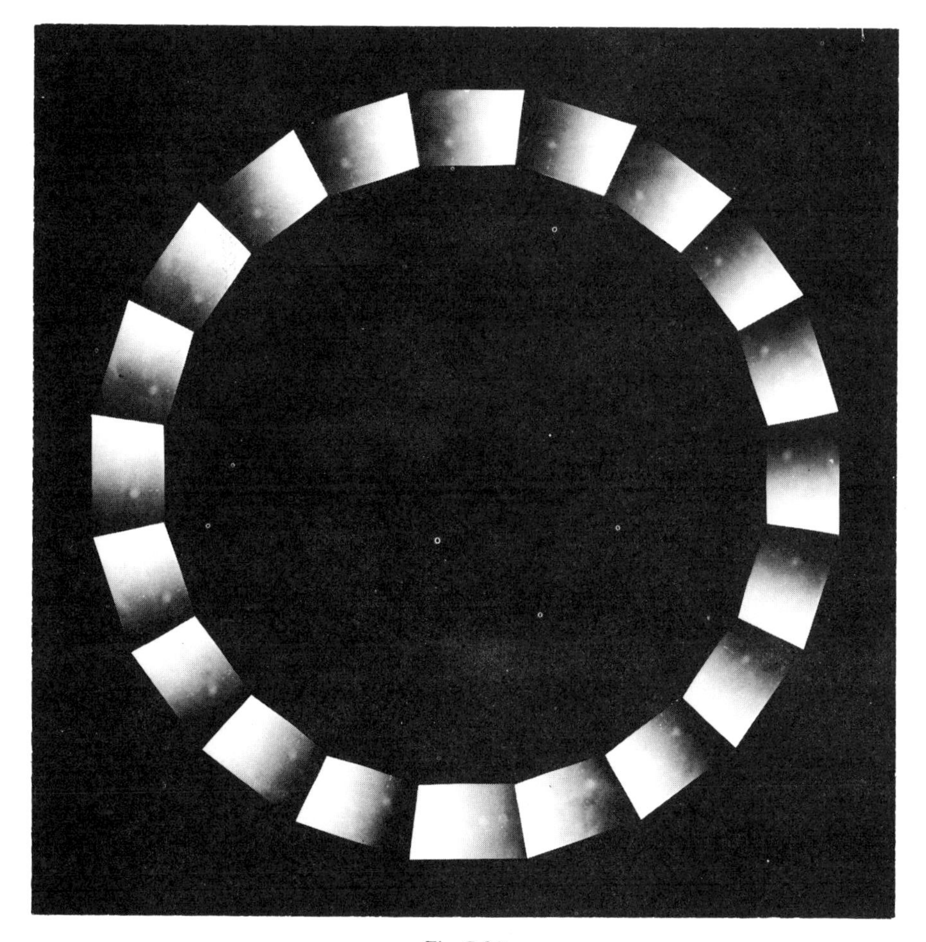

Fig. 7.22

on page 115. If the stripes are graded it is possible to get signals which tell you the direction as well as the speed of rotation. The outputs from a photocell (e.g. a small photodiode) are shown in figure 7.23.

Figure 7.24(a) shows the set-up required to move something in a straight line slowly, and (b) the set-up to move something rapidly.

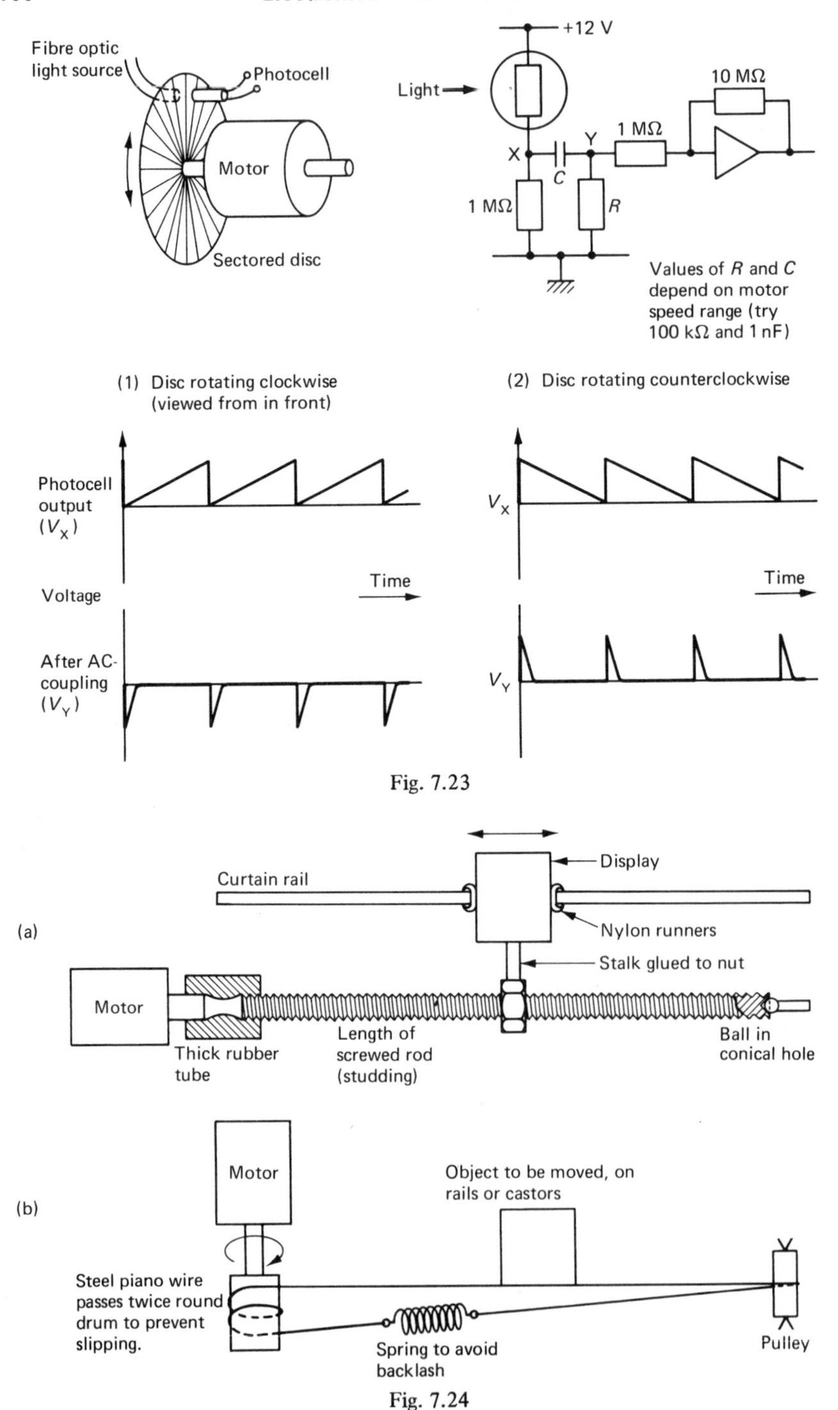


Fig. 7.23

Fig. 7.24

## 7.2 PROGRAMMED STIMULUS PRESENTATION

There are various ways of organising a complex stimulus situation, or in other words of presenting a mixture of visual, auditory, olfactory etc. stimuli in such a way that the times at which each element of the stimulus begins and ends can be pre-set.

### Process Timers

The simplest arrangement is to control each stimulus element and each gap between stimulus elements with a process timer, and to start each clock with a signal from the one controlling the preceding element. To take an example, an experiment studying the interaction of light and olfactory stimuli in an insect might require the sort of stimulus presentation scheme outlined in figure 7.25.

Briefly depressing the two push buttons (preferably mechanically connected) starts timers 1 and 2 operating. Timer 1 has the shutter across its clutch coil, so the lamp comes on at once and stays on for 35 seconds. Timer 2 causes nothing to happen for 25 seconds, but at the end of this time the contacts marked X close briefly causing the relay to operate. The relay has one of its switches connected in the coil circuit such that if it closes briefly once, it stays closed until this extra circuit is interrupted (figure 7.26). If switch B is closed briefly, the relay contacts close and bypass switch B, which can hence be released without the current through the relay coil being interrupted. To reset open switch A briefly.

The relay in this set-up goes on supplying power to the motor till the plunger is fully depressed and all the odour has been pushed towards the insect. At this stage the plunger operates a microswitch which stops the motor and starts the timer controlling the extractor fan, whose function is to ensure quick removal of the odour. After one minute the fan stops running. To reset the system, it is only necessary to wind the plunger back, releasing the microswitch which limits its travel. This allows timer 3 to reset, timers 1 and 2 having done so at the end of the operations they were controlling.

In this set-up everything runs from the AC mains, which saves on power supplies and avoids the problems that arise when transistor circuits pick up stray impulses from mechanical switches. The disadvantages are that great care must be taken to avoid dangerous wiring, and that mechanical switches eventually wear out.

### Magnetic Tape

An alternative approach is to use a tape recorder as the controller. This is a particularly valuable technique for experiments with human subjects because verbal instructions can be recorded on one channel of a stereo tape recorder whose second channel carries the signals controlling the stimuli. Each stimulus source

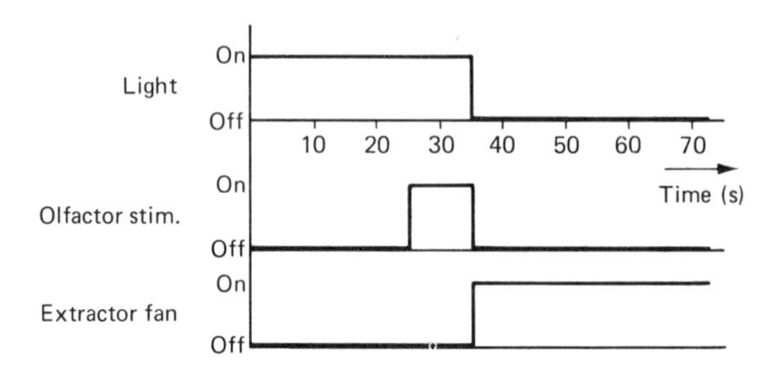


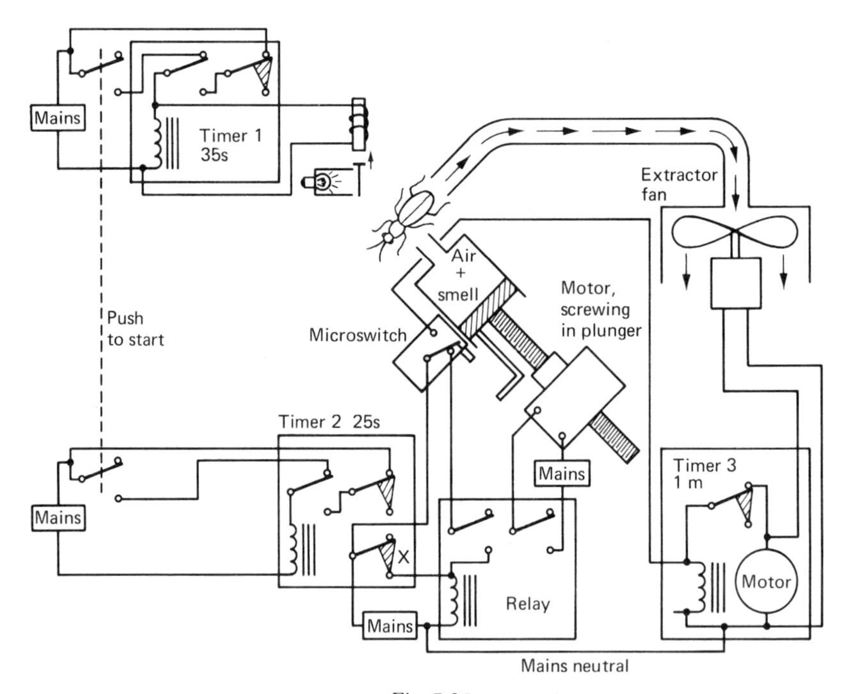


Fig. 7.25

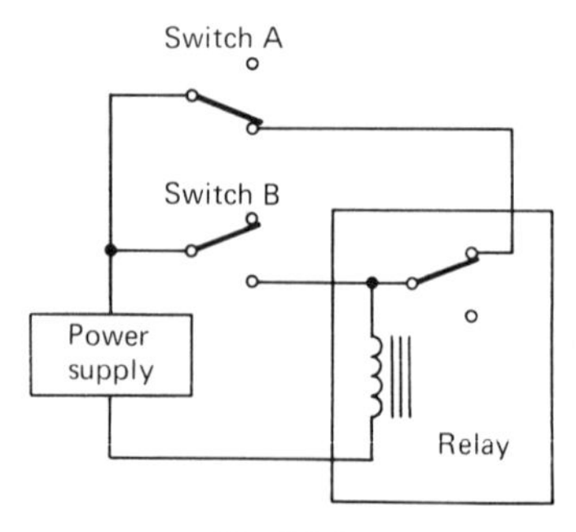


Fig. 7.26

is switched on and off by an integrated-circuit frequency-sensitive switch, a device whose output is zero volts unless a signal of specified frequency (falling within a narrow bandwidth) is present at its input, when the output moves to + 5 volts. The output drives a relay via a transistor switch to control mains-operated equipment, or actuates a transistor switch controlling low voltage DC supplies directly.

Each stimulus source is allotted a different frequency (by suitable adjustment of pre-set potentiometers connected to the integrated circuits) and the controller tape is made by recording a signal from a sine-wave oscillator set to the relevant frequency during each of the time periods in which the stimulus source in question is to be switched on. Two or more signals can overlap without affecting the working of the system.

It is theoretically possible for the frequency-sensitive switches to sort out a great many signals all from the same tape, but problems can arise in the initial recording process. There is no difficulty if a lot of oscillators are available, for then all the signals can be recorded simultaneously. The oscillator outputs are mixed together with an adder (see chapter 4) and the adder output is recorded on tape. Each oscillator is controlled by an operator with a stopwatch and a timetable, who switches it on and off appropriately. However, if only one oscillator is available it is necessary to record each frequency separately using a multiple superimposition technique. To superimpose on something previously recorded, the tape recorder's erase heads are switched off and the new signal recorded on top of the old. The trouble is that even if both signals are recorded at the same level initially, the newer signal tends to predominate markedly when the tape is played back. In general the latest addition supplies about half the total volume, and everything previously recorded the other half. After four or five superimpositions the signals first recorded are beginning to get lost in the noise.

This stimulus presentation system appears at first sight to lend itself to combination with a magnetic tape datalogging system, using different channels of a single tape recorder to control the stimuli and record the responses. Unfortunately most tape recorders require extensive modifications if they are to record on one channel and playback on another. Not only are the multiway switches worked by the 'record' button complex and difficult to sort out, but often bits of electronics have dual functions and will need duplication for simultaneous recording and playback. Two tape recorders cannot be expected to preserve the exact temporal relationships between stimuli and responses. One solution to this problem is to have a datalogger tape recorder with sufficient channels for one to be used to re-record the stimulus control tape as the experiment proceeds.

### Moving Film

For experiments which run for long time periods and for which stimulus control correct to approximately five minutes is adequate, a moving film control system will suffice (figure 7.27). To make a given stimulus source work, stick a piece of black masking tape on an appropriate place on the clear plastic film.

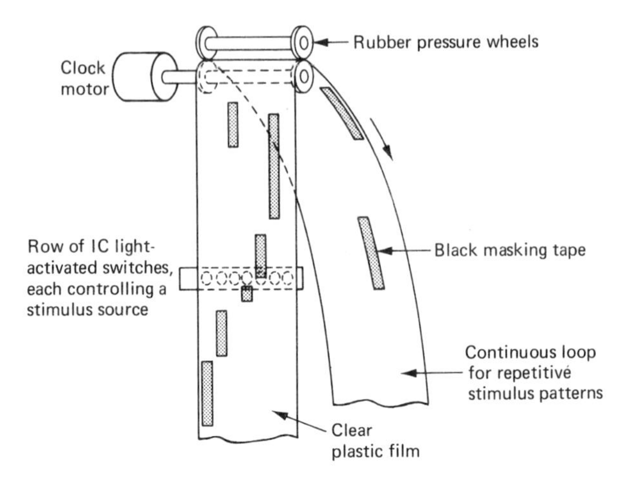


Fig. 7.27

**Paper Tape Readers**

A refinement of the last method is to use computer punched-paper tape instead of the clear plastic film. The simple-minded way of doing this is to use the tape to control only four stimulus sources, with a simple code, e.g. a hole in position 1 means 'start stimulus source 1', a hole in position 2 means 'stop stimulus source 1' etc.

The punched-paper tape is played back using a machine called a tape reader, which has a row of eight photocells to detect whether or not there are holes in the tape, and a motor which jerks the tape forward each time it receives a command signal till the next row of holes is over the photocell array.

To perform an experiment each pair of photocells is connected to a separate stimulus source and the paper tape advanced a notch every time an interval timer resets. This can be as fast as several hundred times a second if necessary. The great advantage of this technique is that the paper tapes can be cut by a computer. This means that you can substitute typing the timetable into the computer and letting an appropriate program generate the paper tapes for spending hours with scissors and masking tape. The disadvantage is that stimuli can only start or stop at the instant the tape jerks forward in the reader. The limitation to four channels is an artificial one. Provided you are prepared to make the digital electronics required to convert a pattern of voltages representing a binary number into a pulse along a wire uniquely representing that number, each row of eight holes in the tape can specify that a pulse is output along one of 256 separate wires.

To take an example, consider an experiment with 125 stimulus sources, and a time-scale which made changing the stimuli roughly every five minutes adequate.

Each stimulus source would be connected to two wires, one to start and one to stop it, and would hence be represented inside the computer by two eight bit binary numbers. A timer would tell the tape reader to start reading at the end of a five-minute gap. The tape contains a list of the numbers representing each of the stimulus sources which need changing. Each number occupies a whole row of eight holes, so it is necessary to read a series of rows of holes as quickly as possible, acting at once on each. One number would be used to mean 'end of batch' and its output wire would stop the reader when a pulse came along it.

## 7.3 INTERACTIVE STIMULUS PRESENTATION

This is a form of stimulus control in which the stimulus situation depends at any instant on previous responses by the system being investigated.

For example, on-line digital computers have been used to plot the visual receptive fields of nerve cells in visual areas of the cat's brain. To plot a receptive field a micro-electrode is placed in a cell in the brain, and attempts are made to change the rate at which this cell fires by altering the visual display in front of the animal's eyes. A typical result would be that a given cell gives its maximum response if a line of certain length at a fixed angle to the horizontal moves at a certain speed across the visual field. This sort of complicated pattern takes many hours of trial-and-error testing to discover, and there is a limit to how long you can keep a micropipette stuck into a cell without the cell dying.

The computer produces the visual display by controlling patterns on a large cathode

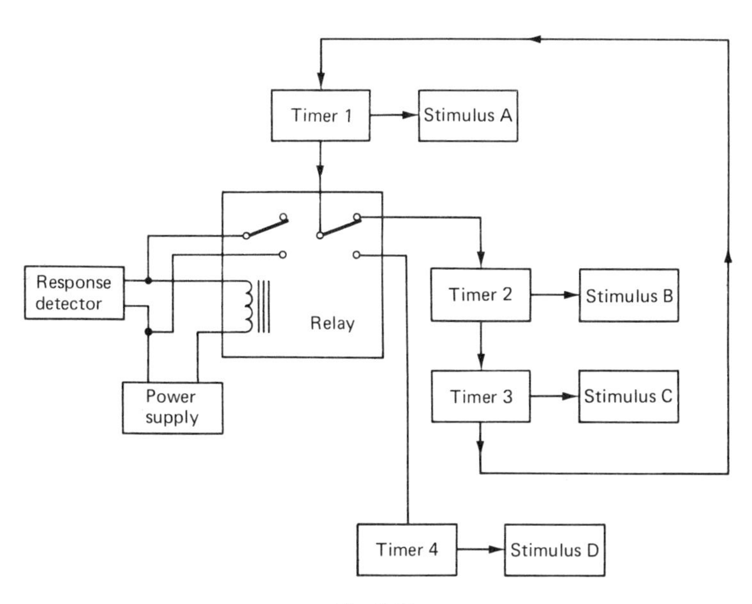


Fig. 7.28

ray tube in front of the cat. Each spike from the cell being recorded from is input to the computer, which alters its display in such a way as to maximise the rate of firing of the cell. The programs required to perform this optimisation procedure (often called hill climbing) are complex and require a large computer. Basically what happens is that the program represents the stimulus as a point in a multidimensional space, with one dimension for each of the ways in which the stimulus can be altered. It tentatively moves the spot a little in all possible directions till the nerve cell gives some response, then it varies the scales of the various dimensions to maximise the response.

The main problems which arise in doing this kind of work are administrative ones. It is extremely difficult to find any one prepared to allow you to feed electrical signals directly into a large computer, and the large amounts of machine time required do not fit very well into most time-sharing systems.

On a much simpler level it is possible to introduce some elements of interactive control into the sort of electromechanical stimulus control shown in (figure 7.28). This is the kind of technique some teaching machines use; you simply build two (or more) different stimulus control systems, the decision as to which one to use depending on whether a given response event occurs. A repetitive stimulus pattern continues till a response occurs, when the relay changes over and stays changed, allowing the 'branch' to be activated next time round the loop.

# 8 Pulse Circuits

## 8.1 THE DELAY CIRCUIT OR MONOSTABLE

### Function

This circuit is the basis of electronic timers. Left to itself it never leaves its stable state (figure 8.1). The output remains indefinitely very close to zero volts. If you connect the input briefly to a positive supply the output immediately shoots up to plus + 12 volts, stays there for a pre-determined period of time, and then returns sharply to zero volts. The length of the predetermined time interval is set only by the values of the resistor and capacitor marked $R$ and $C$

$$t = 0.7\,RC$$

where $t$ is measured in milliseconds, $R$ in kilohms and $C$ in microfarads. Hence for the circuit shown $t = 0.7 \times 50 \times 0.1 = 3.5$ milliseconds. The circuit works best for times between 100 milliseconds and 100 microseconds, but, by careful design, it is possible to make versions which operate for periods up to 1 minute, and down to 100 nanoseconds.

### Operation

The actual working of this circuit is not nearly as straight-forward as you would expect from the simple appearance of the diagram, so do not expect to understand the following account at a first reading. Most people find they have to make several monostables using rule-of-thumb design before they become really clear about the way they work.

We begin with the stable state. The right-hand transistor receives a sufficient current flow into its base via the 50 kΩ resistor to keep it firmly 'turned on'. In other words, the collector current flowing through the right-hand 2 kΩ resistor is large enough to bring point Z very close to zero volts. Hence only a very small current flows along the 20 kΩ resistor from Z to the base of the left-hand

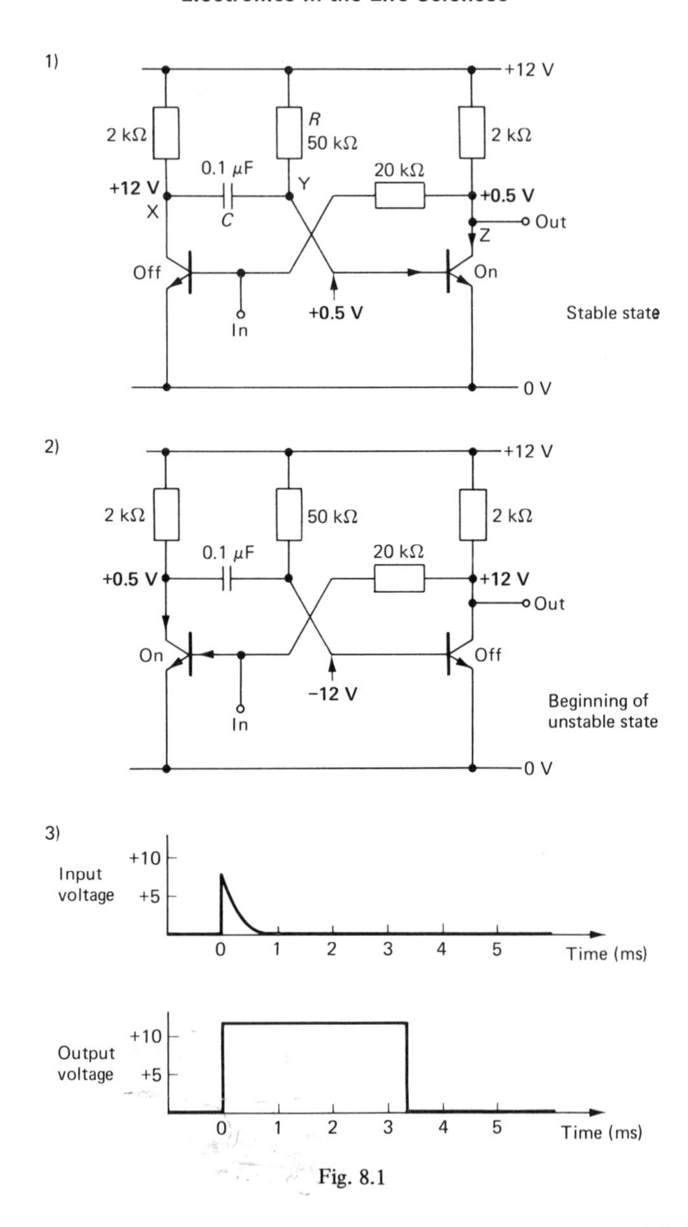


Fig. 8.1

transistor, which is thus turned off (i.e. no current flows through the left-hand 2 kΩ resistor, and X is at + 12 volts; 12 volts are thus maintained across the capacitor). Because of the low resistance to positive-going currents between the base and the emitter of an NPN transistor all the current flowing down the 50 kΩ resistor passes through the right-hand transistor to the zero volt rail, and none flows into the capacitor. In this context, it is worth remembering the relationship between transistor and diode symbols (figure 8.2). So, in the case of the right-hand transistor, 'the base-emitter diode' is forward biassed, and current flows through the transistor.

To make the circuit change to its unstable state we connect a positive voltage briefly to the input ('apply a short positive-going pulse'). This flow of current into the base of the left-hand transistor causes a larger flow down its collector resistor,

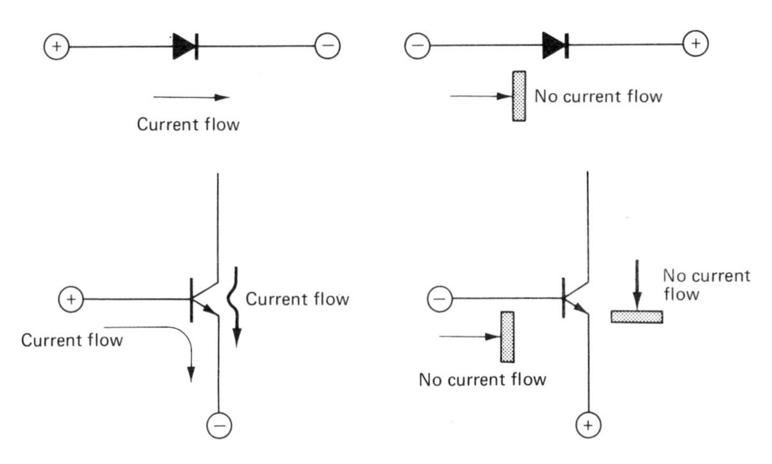


Fig. 8.2

bringing X a little away from the + 12 volts towards zero volts, that is the input pulse starts to turn the left-hand transistor on. These effects are immediately transmitted to the right-hand transistor, for when X, connected to the left-hand side of the capacitor, moves from + 12 to + 11 volts, point Y, connected to the right-hand side of the capacitor, will move from + 0.5 volt to − 0.5 volt. A change in this direction at once begins to turn the right-hand transistor off by reducing its base current, and Z begins to move towards + 12 volts. This movement is in its turn fed back to the left-hand transistor through the 20 kΩ resistor joining its base to Z, and the left-hand transistor is turned on even further. A vicious circle of this sort reaches the unstable state extremely quickly, with the left-hand transistor now fully turned on, and the right-hand transistor turned off. Notice that during the whole of this brief process the voltage between the two ends of the capacitor remains at 11.5 volts, so that by the time the unstable state is reached, Y is down near − 12 volts.

Now that the right-hand side transistor has a backward biassed base-emitter diode, and no current can flow into it, the whole of the current flowing down the 50 kΩ resistor flows into the capacitor, so the voltage at Y slowly moves up towards zero volts. Until it gets there nothing at all happens at the output which stays up at + 12 volts. However as soon as Y does get above zero volts the right-hand transistor's base-emitter diode starts to conduct again, and all the current begins onee again to flow into the transistor, tending to turn it on. This, of course, activates the crossed feedback loops, and we come clattering back down into the stable state, with the left-hand transistor on and the right-hand one off. During this process X returns to + 12 vol just as quickly as the capacitor can be recharged by the left-hand 2 kΩ resistor, for, by the end of the unstable period, both sides of the capacitor were at zero volts.

So far a circuit involving NPN transistors and outputting positive-going pulses has been considered. As usual it is possible to obtain negative-going output pulses for negative-going input pulses by substituting PNP transistors and connecting the top rail to − 12 volts.

## Design

Normal monostable circuits have two slight additions to the circuit shown in figure 8.1 (figure 8.3).

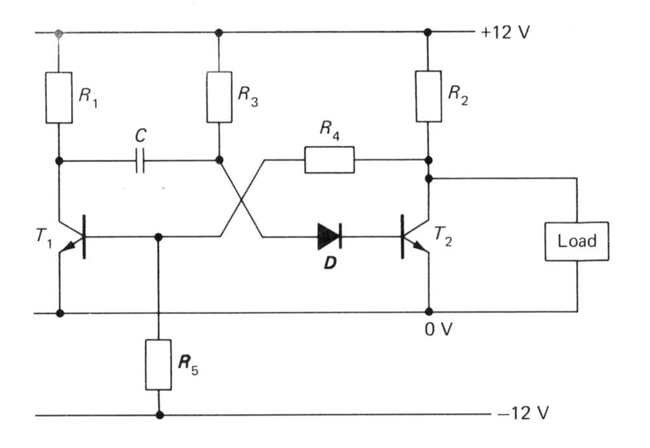


for pnp transistors:

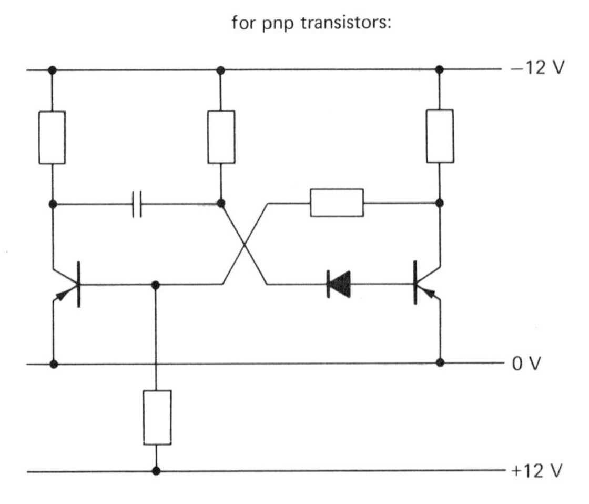


Fig. 8.3

The diode $D$ is added because most silicon transistors do not like reverse base-emitter voltages of greater than 5 volts – the purpose of the diode is hence just to strengthen the transistor's internal diode described above. $R_5$ is to prevent the left-hand transistor coming on unintentionally. When choosing values:

$R_1 = R_2$: keep both fairly low, say in the range 1.5 to 5 kΩ, because (1) $R_2$ is in effect in series with the load, and hence the actual voltage applied to the load will be divided in the ratio of $R_2$ to $R_{load}$, that is with $R_2$ = 2 kΩ and $R_{load}$ = 2 kΩ, you would only get 6 volts applied to the load. (2) $C$ has to recharge through $R_1$, and if this is large the return to the stable state will be slowed down.

$R_5$: normally 50 or 100 kΩ. If your circuit goes off continuously on its own, try reducing this value; if it requires enormous pulses to go off at all, try increasing it, or leaving it out altogether.

$R_4$: should be between 10 times and 20 times $R_1$.

$R_3$: together with $C$ this sets the time spent in the unstable state. Choose an $R_3$ such that it is less than

$$R_1 \times \left( \frac{\text{gain of right-hand transistor}}{2} \right)$$

$C$: as capacitors get drastically more expensive as they get bigger, it is best to keep $R_3$ as big as possible, and hence $C$ as small as possible. Avoid electrolytic capacitors if you can (they leak) but if you do have to use them to get them the right way round. Figure 8.4 shows the circuits for these capacitors with (a) NPN transistors and (b) PNP transistors. Long intervals can be achieved by using a large $R$ and an extra transistor (figure 4.4(c)).

$T_1$ and $T_2$: except for the fastest monostables there is no need to use expensive high-speed switching transistors, since any audio-type transistor will work perfectly well (e.g. 2N3704, 2N3702). $T_2$ should have a high gain. For very long periods it is best to use two transistors connected as shown in order to multiply the current gains together. For example if both transistors have gain of 100, the pair behave as a single transistor of gain 10 000, and hence a very large $R_3$ may be used.

$D$: any signal diode which will stand a reverse voltage of 12 volts will serve. (e.g. 1S920)

### Arranging a Suitable Short Input Pulse

It is usual to couple a monostable to the circuit or to the switch driving it with a simple pulse-shaping network (figure 8.5). It is best to keep the coupling capacitor as small as possible, and the value should not normally need to exceed 500 pF unless the input signal deviates very badly from a square wave. For example, the signal shown in figure 8.6(a) will easily trigger a monostable with a 220 pF capacitor in its input network, while the signal in figure 8.6(b) may require a 5000 pF capacitor.

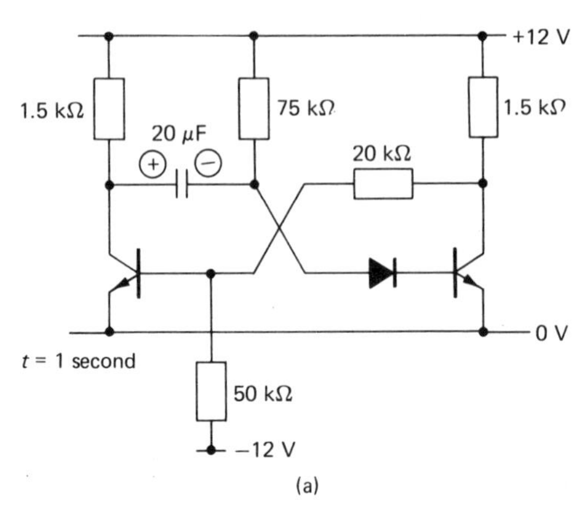

(a)

The capacitor goes the other way round for PNP transistors:

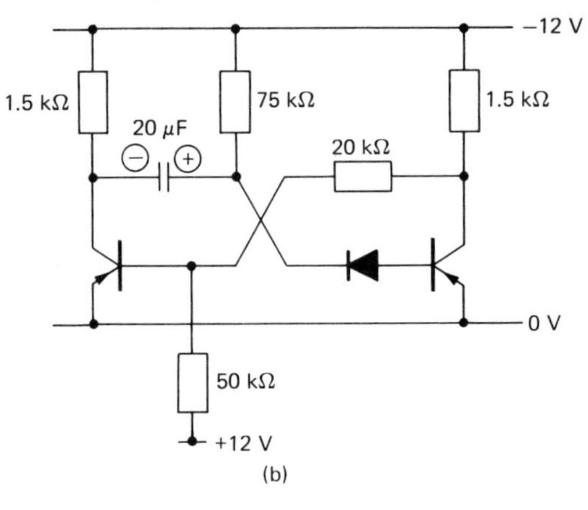

(b)

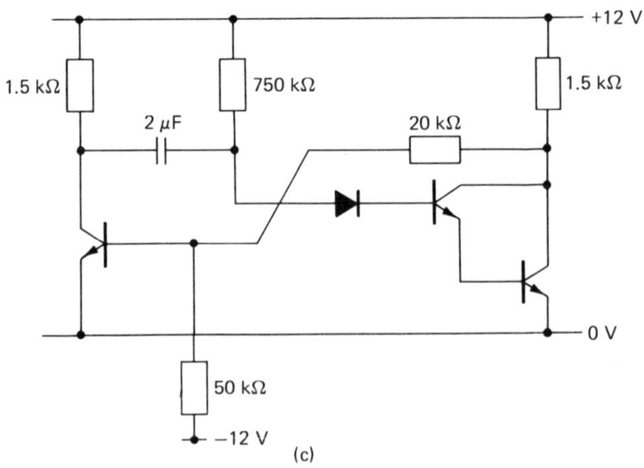

(c)

Fig. 8.4

If you need to drive an NPN transistor monostable with a negative-going pulse, it is possible to do so by inputting it to the right-hand transistor, which begins to be turned off by the input pulses. Hence we have the four possible cases shown in figure 8.7. It is possible to prevent a monostable from operating during some specified

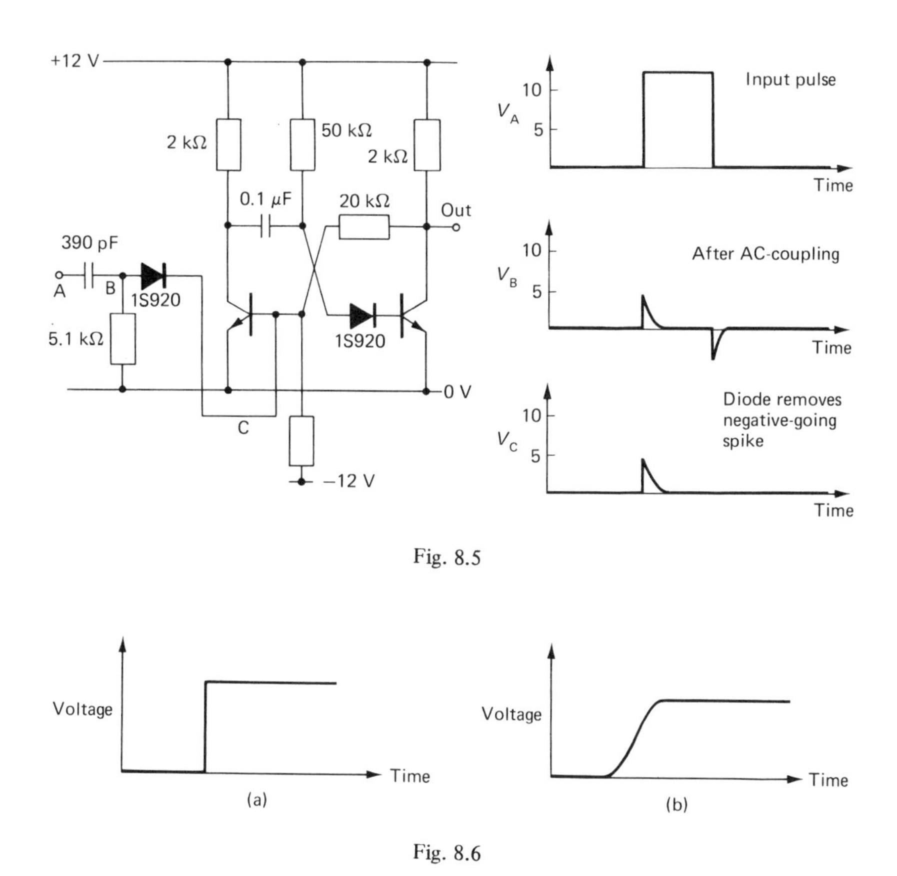


Fig. 8.5

Fig. 8.6

time period by biassing the diode in the input network backwards (figure 8.8). If $R_{in}$ is connected to the $-12$ rail (say, by another monostable) the input network is blocked to incoming pulses during the time this connection is maintained. This paralysing input can be useful, if, for instance, you are using the monostable to drive the counter coils in a printing counter and need to prevent the count reached being altered during the brief period while printing is taking place.

It is possible to arrange a monostable so that it is capable of being triggered from many sources. As shown in figure 8.9, you can use the collector of the left-hand transistor as an auxiliary output of negative-going pulses. The leading edges of these pulses will be as shown, less square than those from the right-hand side because of the necessity to recharge the capacitor during the return to the stable state.

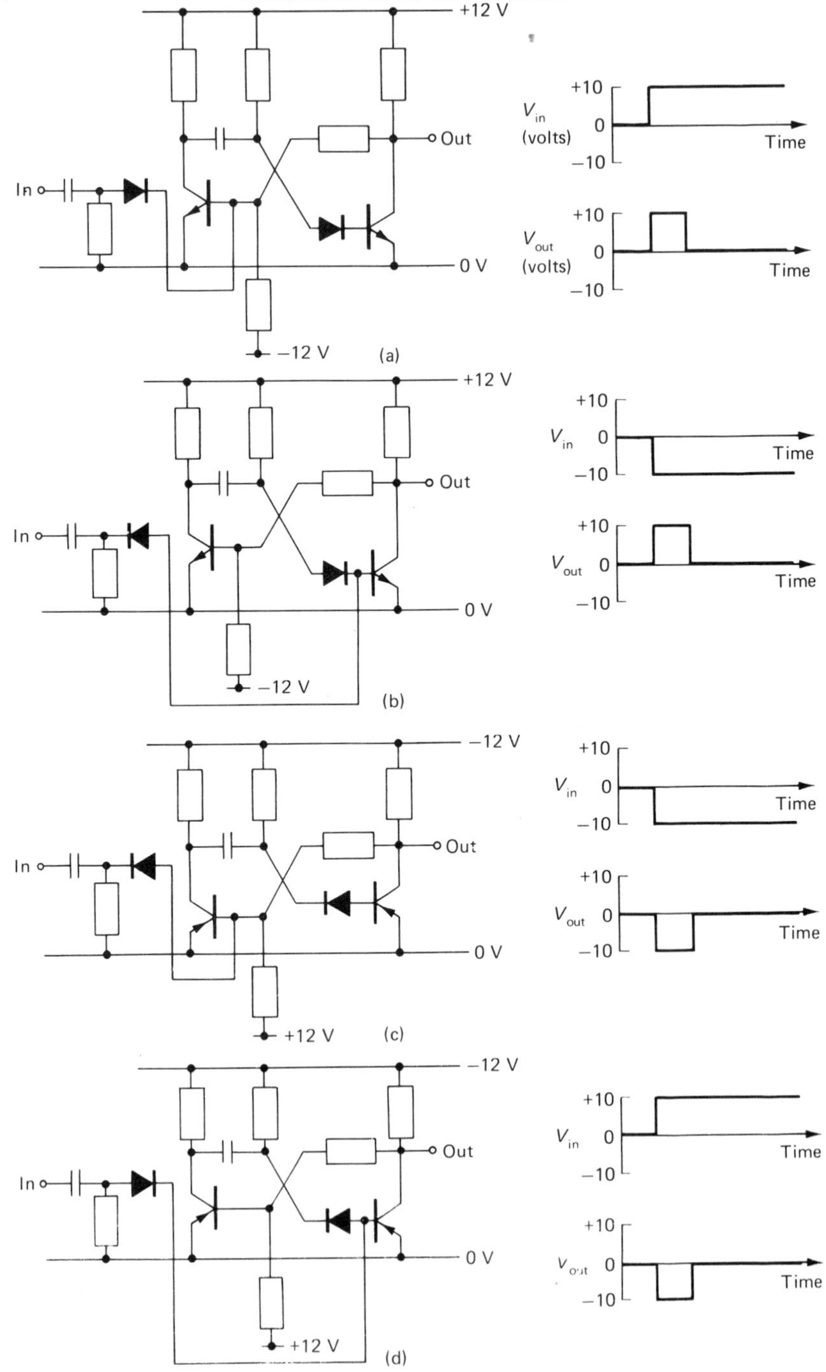
+12 V
Out
In
0 V
−12 V
(a)
$V_{in}$
(volts)
+10
0
−10
Time
$V_{out}$
(volts)
+12 V
Out
In
0 V
−12 V
(b)
−12 V
Out
In
0 V
+12 V
(c)
−12 V
Out
In
0 V
+12 V
(d)

Fig. 8.7

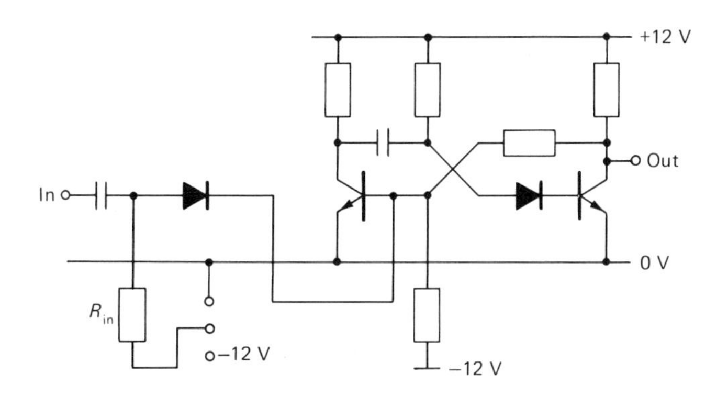
+12 V
Out
In
0 V
$R_{in}$
−12 V
−12 V

Fig. 8.8

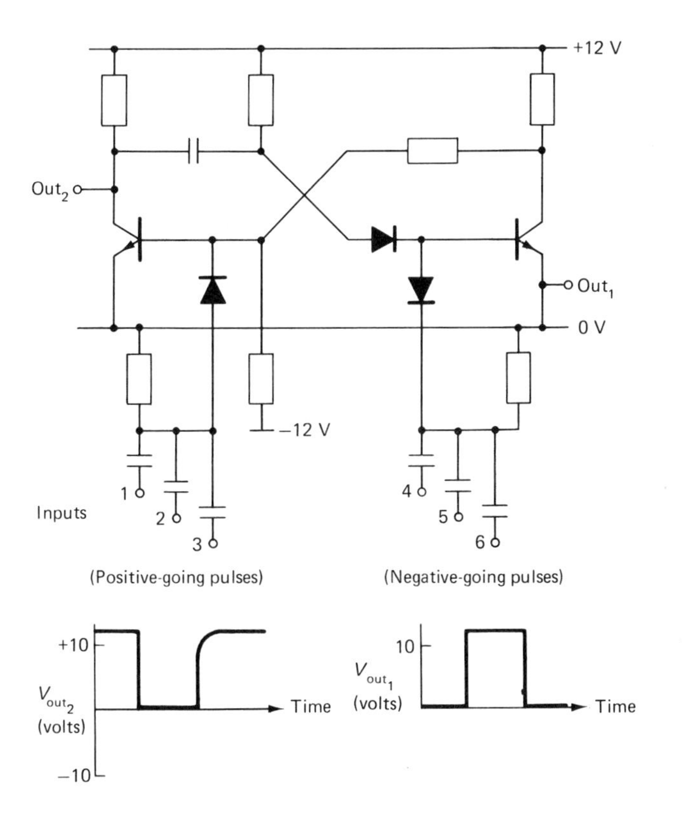
+12 V
$Out_2$
$Out_1$
0 V
−12 V
Inputs
1
2
3
4
5
6
(Positive-going pulses)
(Negative-going pulses)
+10
$V_{out_2}$ (volts)
−10
Time
10
$V_{out_1}$ (volts)
Time

Fig. 8.9

## 8.2 THE MULTIVIBRATOR

### Function

This circuit is an oscillator; it requires no special inputs, and, as soon as the power supply is applied, outputs a continuing square-wave of predetermined frequency. It is very useful, for example, for driving the transistor switches alternately sampling two input channels in a data multiplexer, or for supplying a carrier frequency for an amplitude-modulated recording system (figure 8.10).

The period of the oscillation is set by the pairs of components marked $R$ and $C$. In this case, $t = 1.4\ RC$, where $t$ is the time in milliseconds for one complete cycle to occur, $R$ is measured in kilohms, and $C$ in microfarads.

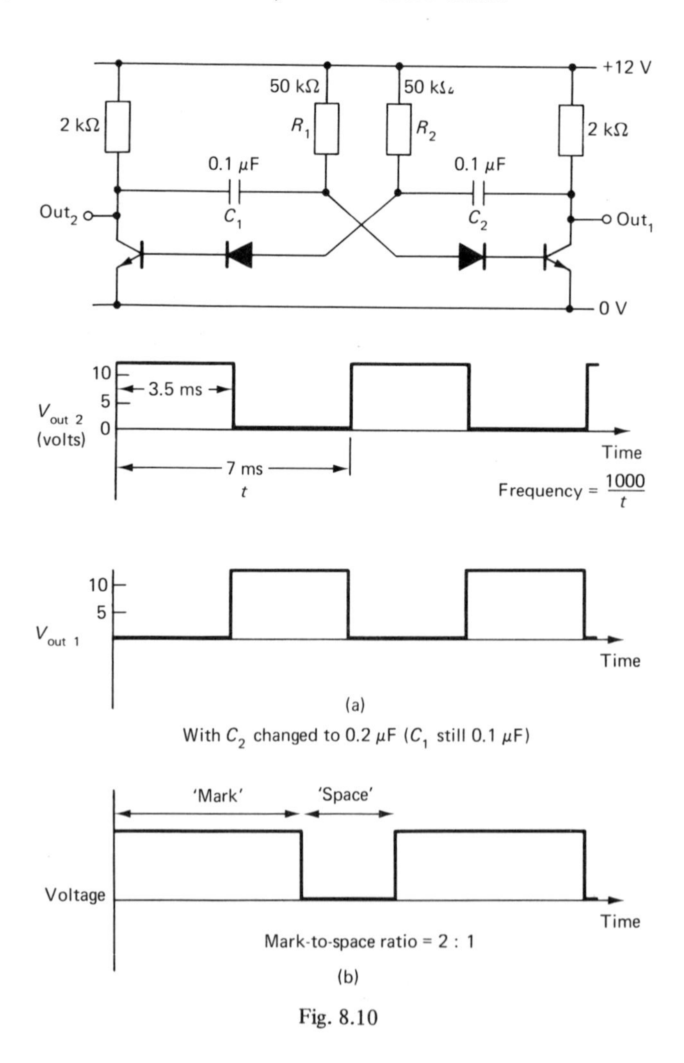


Fig. 8.10

## Operation

It will be clear from the circuit diagram that this is just two left-hand sides of monostables stuck back to back so that they repetitively trigger each other. Every change of state follows the pattern described above for the return of the monostable from its unstable to its stable state, and each half-monostable takes it in turn to behave as if it were in the stable state, the other half simultaneously going through the unstable process of bringing its capacitor up from − 12 volts to a point where switching can occur.

It is difficult to see what starts a multivibrator off when you switch the power on, for there is nothing to prevent it settling down with both sides in the same state, particularly if it is exactly symmetrical, as shown. However, in practice this very rarely seems to present a problem, for most actual multivibrators never fail to start, and indeed can be relied on to start always in the same one of the two possible states. There is no reason why $R_1$ should always equal $R_2$ and $C_1$ equal $C_2$. If they do not, a multivibrator with a mark-to-space ratio not equal to one will result.

## Design

Most of the considerations are similar to those for monostables (figure 8.11).

$R_1$ and $R_2$ : should be equal and small: 1.5 kΩ to 5 kΩ.

$R_3$ and $R_4$: should not be more than (transistor gain)/$2R_1$.

$C_1$ and $C_2$ : should be kept as small as possible, both on grounds of cost and also because of the bad effects of the recharge of the capacitors through the collector resistors on one of the reset times. In fact, looking at the waveform from the right-hand side (figure 8.12), there is always going to be some asymmetry in the rise times obtained from this circuit.

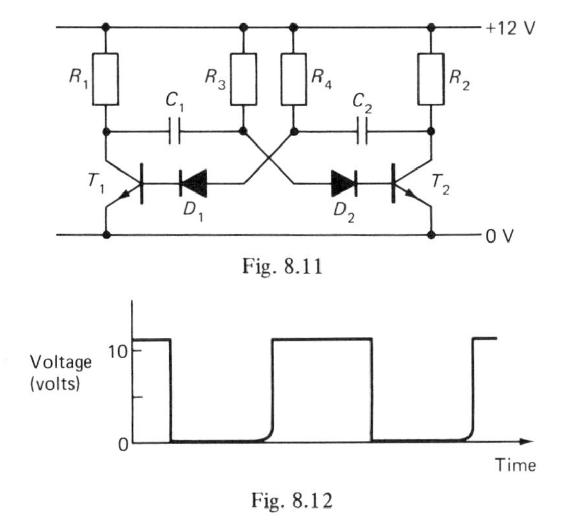


Fig. 8.11



Fig. 8.12

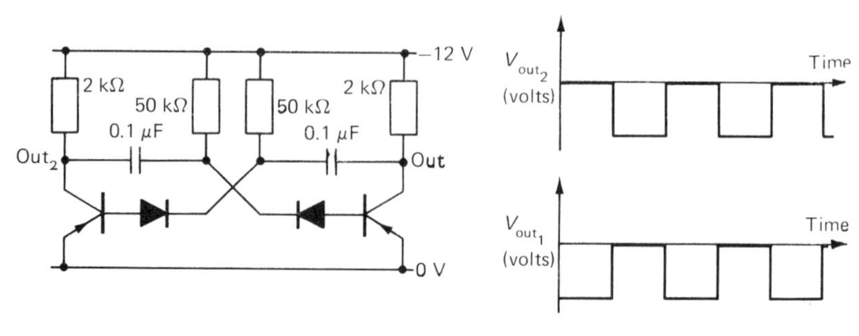


Fig. 8.13

Diodes and transistors: exactly as for a monostable. As usual, PNP transistors can be used (figure 8.13).

## 8.3 THE BISTABLE (FLIP-FLOP)

### Function

This circuit is the primary building block of the digital computer. As usual in pulse circuits the transistors are connected so that if one of them is conducting (on) the other is bound to be off (figure 8.14).

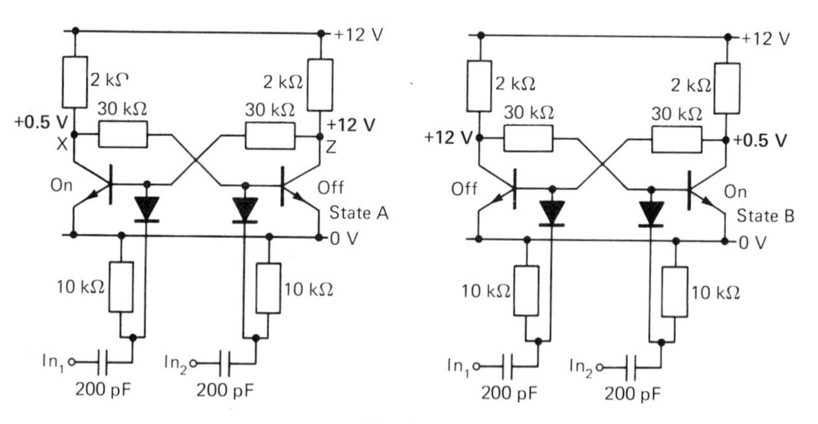


Fig. 8.14

There are hence two possible states for this circuit: either the left-hand transistor is on and the right-hand transistor is off (state A), or else the left-hand transistor is off and the right-hand one on (state B). In this case both of the states are indefinitely stable, provided that the power supply is maintained. State A is transformed into state B by applying a negative-going pulse to input 1, and state B continues until a negative-going pulse is applied to input 2, whereupon the circuit reverts to state A.

This circuit can thus be said to 'remember' a pulse till it is told to 'forget' it.

**Operation**

Considering state A: a negative-going pulse applied to input 1 will tend to turn the left-hand transistor off a little. This means that point X begins to move towards + 12 volts, and that the base current of the right-hand transistor begins to increase as current begins to flow down the 30 kΩ resistor joining its base to X. Thus the right-hand transistor begins to come on, that is, X begins to move from + 12 towards zero volts. This of course has the effect of turning the left-hand transistor still further off, and our crossed feedback loops rapidly take the circuit through to state B. As everything is totally symmetrical, the reverse change, occurring only when a negative-going pulse is applied to input 2 (further pulses applied to input 1 have no effect) follows exactly the same course as described above. It is usual to trigger bistables by turning the 'on' transistor off, but in fact the alternative strategy of turning the 'off' transistor on a little with a positive-going pulse would work equally well.

One very common arrangement of the triggering networks is that shown in figure 8.15. This is the circuit represented by a block diagram labelled 'bistable

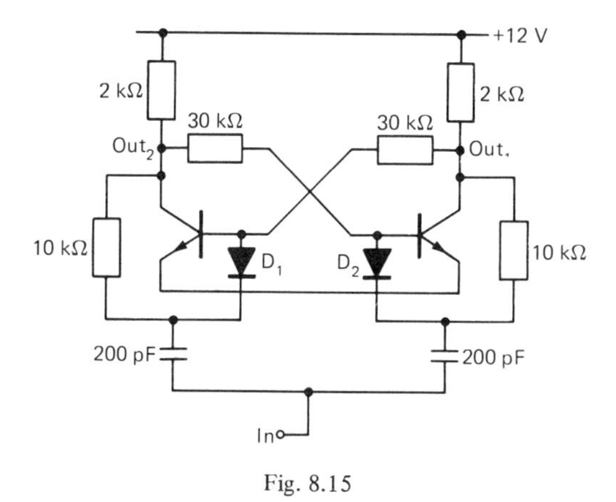


Fig. 8.15

memory' in chapters 6 and 7. As noted there it operates like a retractable ball-point pen—each impulse changes the circuit from whichever state it happens to be in when the pulse arrives to the opposite state. If the left-hand transistor is off to begin with, point X will be at + 12 volts, and diode $D_1$ will be biassed backwards through the 10 kΩ resistor connecting it to X. So an incoming pulse is directed through diode $D_2$ (the other end of its 10 kΩ resistor, point Z, being nearly at zero volts) and serves to change the state of the circuit, turning the left-hand transistor on, and the right-hand transistor off. Thus, when the next pulse arrives, diode $D_2$ is now biassed backwards, and the pulse is directed to the left hand-transistor.

Bistables sometimes tend to be unstable and flip states spontaneously. The cure is to connect resistors from the transistor bases to the − 12 volt rail to ensure that a transistor meant to be 'off' stays that way (figure 8.16).

At the beginning of the operation of a complex circuit it is often necessary to ensure that all the bistables involved are in the same state, generally state A (out

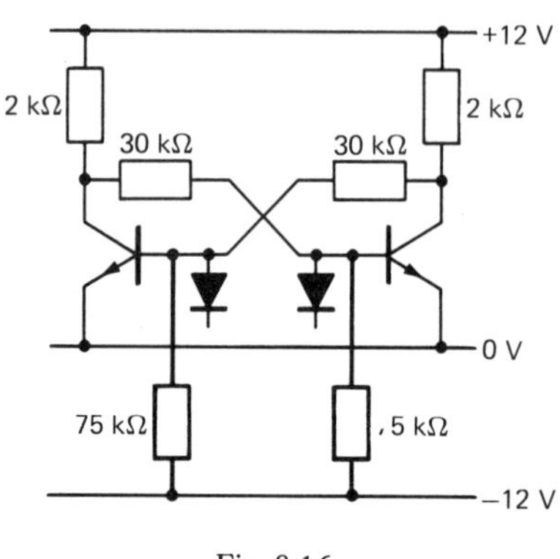


Fig. 8.16

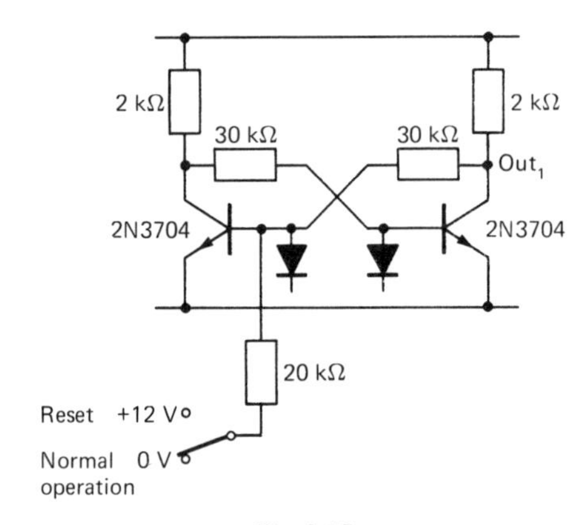


Fig. 8.17

1 = 0 volts). This does not happen automatically when the power supply is connected, because there is no mechanism in a bistable to make either state more likely to occur initially than the other. Hence the need for a reset line, connected via a resistor to the base of the left-hand transistor of each bistable. This line, when connected to + 12 volts, will turn on any of the left-hand transistors which are off, and thus ensure that all the bistables are in state A (figure 8.17).

# Manufacturers and Suppliers

Inclusion in this list does not indicate that the product mentioned is in any sense a 'best buy', but merely that I have used it at some stage, and that it did work reasonably well.

**Components**

It is difficult to obtain components in small quantities; watch out for long delays and swingeing minimum order charges. The only British firm which actually seems to like small orders is RS Components.

| Product | Supplier |
|---|---|
| Electronics boxes<br>Capacitors<br>Circuit board<br>Connectors<br>Counter decades<br>Diodes and diode bridges<br>Heat sinks<br>IC op amps and comparators<br>Potentiometers<br>Power transistors<br>Voltage regulators<br>Relays<br>Resistors<br>Thermistors<br>Transformers | RS Components Ltd., 13 Epworth Street, London EC 2 |
| Transistors and FET's | Texas Instruments, Manton Lane, Bedford |
| IC monostables, frequency sensitive switches | Consumer Microcircuits, Rickstones Rd., Witham, Essex |

**Test Instruments**

| | |
|---|---|
| Digital voltmeter, oscilloscope | Advance Electronics Ltd., Raynham Rd., Bishop's Stortford, Herts. |
| Multimeter | Avo Ltd., Archcliffe Rd., Dover, Kent |
| Oscillator | Levell Electronics Ltd., Park Rd., High Barnett, Herts. |
| Storage oscilloscope | Telequipment, 313 Chase Rd., Southgate, London N 14 |

**Data Recording Equipment**

Counters ('post office'), and other surplus items, e.g. small motors, see advertisements in back of current '*Wireless World*'.

| | |
|---|---|
| Printing counter | English Numbering Machines Ltd., Queensway, Enfield, Middlesex |
| Tape punch | Westrex Ltd., 152 Coles Green Rd., London NW 2 |
| Stereo tape recorder | Revox, C. E. Hammond, Church Street, London NW 2 |
| 4 channel, FM tape recorder | Fenlow Electronics Ltd., Jessamy Rd., Weybridge, Surrey |

**Physiological**

| | |
|---|---|
| Electrodes (readymade) | Frederick Haer Europe, Pangbourne, Reading |
| Preamplifiers, A–D converters | Ancom Ltd., Devonshire Street, Cheltenham, Gloucestershire |

**Optical**

| | |
|---|---|
| Interference filters, wedges, glass fibre optics | Barr and Stroud Ltd., Anniesland, Glasgow |
| Plastic fibre optics (and CdS and Se photocells) | Proops Brothers Ltd., 53 Bayham Street, London NW 1 |
| Quartz fibre optics | Jenaer Glaswerk, H. V. Skan Ltd., 425 Stratford Rd., Shirley, Solihull, Warwickshire |

**Opto-electronic**

| | |
|---|---|
| Solid state lamps | MCP Electronics, Alperton, Wembley, Middlesex |
| Photometric diodes | United Detector Technology, Techmation Ltd., 58 Edgeware Way, Edgeware, Middlesex |
| Light-activated switch | RS Components |
| Photomultipliers, vacuum photocells<br>BPY68 photodiodes | Mullard Ltd., Torrington Place, London WC 1 |
| Photomultiplier Magnetic shields | Magnetic Shields Ltd., Headcorn Rd., Staplehurst, Tonbridge, Kent |
| EHT power supplies | Brandenburg Ltd., 939 London Rd., Thornton Heath, Surrey |
| Vacuum thermopile | Hilger and Watts Ltd., 98 St. Pancras Way, London WC 1 |

**Mechanical**

| | |
|---|---|
| Process timers | IMO Precision Controls, Ltd., 313 Edgeware Rd., London W 2 |
| Stepping motors | Impex Electrical Ltd., Market Rd., Richmond, Surrey |
| Velodyne motor–tachogenerator sets | Evershed and Vignoles Ltd., Acton Lane, London W 4 |

**Miscellaneous**

| | |
|---|---|
| Variable autotransformer 'variac' | General Radio Company Ltd., Bourne End, Bucks. |
| Thermocouple Cable (copper, Constantan) | Stirling Cable Co. Ltd., Bath Rd., Aldermaston, Reading, Berks. |

A moderately useful (but expensive) directory to firms and products in this general field is the IEA purchasing directory, published by Morgan Grampian Ltd., 28 Essex Street, London WC 2.

# Glossary

This includes some technical and jargon terms avoided in the text. Words in the main index are in bold type. '=' means 'is a synonym for'.

ACTOGRAPH = **activity detector**
ANALOGUE systems represent things by the sizes of voltages or currents.
BANDWIDTH is the capacity of a communication channel.
BISTABLE LATCH is a circuit for remembering a binary number. (One is described on p. 112.)
CARRIER frequency wave has a signal impressed on it by, for example, a radio transmitter.
DEMODULATOR reconstitutes a coded signal.
DIGITAL systems represent things by patterns of dots and dashes.
EARTH LOOP occurs when a piece of equipment has two separate connections to **earth**. It causes **hum.**
ELECTROMETER AMPLIFIER has an enormous **input impedance** (e.g. that shown in figure 4.53, p. 30).
ELECTRON GUN shown in figure 2.22, p. 28.
EVENT RECORDER is a **pen recorder** whose pens give small blips to indicate when an event occurs.
FARADAY CAGE. An example of this is the steel wire cage described on p. 81.
FEEDBACK occurs when something's output affects its input.
FILTER passes signals only within a certain range of frequencies (e.g. **high-pass filter, low-pass filter**).
FLIP-FLOP = **bistable.**
FLOATING outputs have a voltage difference between two wires, either of which can be connected to ground or any convenient voltage level.
FREQUENCY HALVER = **bistable memory**
HARD OVER means that an amplifier output has reached its maximum voltage and further increases in input will have no effect.

HYSTERESIS. An example of this is the gap between the temperature at which a thermostat switches on and that at which it switches off.
IMPEDANCE is AC resistance.
LIGHT EMITTING DIODE = **solid state lamp**.
MISMATCH occurs when one circuit tries to draw too much current from another
MODULATOR encodes a signal for recording or transmission.
MULTIPLEXING is sharing a single communication channel between lots of signals.
NOISE is random unwanted signal (e.g. VHF radio between stations).
ON LINE means being connected directly to a computer, which deals at once with any input data.
ONE SHOT MULTIVIBRATOR = **monostable.**
PERFORATOR = **tape punch.**
PHASE SHIFT has happened to the output voltage in figure 2.15, p. 24.
PHOTOVOLTAIC CELL. An example of this is the **selenium photocell.**
PLUMBICON. A TV camera tube for dim light.
POLYGRAPH = **multichannel pen recorder.**
POST OFFICE COUNTER is a 4 digit **electromagnetic counter.**
POTENTIAL DIVIDER = voltage divider.
POTENTIOMETER = volume control.
PULSE. A brief signal whose shape is as shown in figure 5.10, p. 87.
RASTER. The pattern of lines making up a TV picture.
SAWTOOTH OSCILLATOR. The circuit in figure 5.22, p. 95 is an example.
SERVOCONTROLLED MOTOR = **velodyne.**
SERVOMECHANISM maintains a controlled state of affairs by monitoring its output and automatically making appropriate adjustments.
SHAFT ENCODER. An example of this is the motor speed monitor on p. 167.
SPECTRAL EMISSION CURVES = **colour distribution curves.**
STEERING CIRCUIT = **bistable trigger network.**
TACHOGENERATOR = **velodyne** generator.
THERMAL NOISE, see noise (p. 70).
TRANSDUCER converts change in some other modality (heat, light movement etc.) into an electrical signal.
VARIAC = variable **autotransformer.**
VOLTAGE CONTROLLED OSCILLATOR has an output frequency proportional to an input voltage. See FM modulator circuit, p. 119.
WINDOW DISCRIMINATOR = **pulse height discriminator.**

# Index